Physiology Illustrated

Physiology Illustrated

Olof Lippold and Barbara Cogdell

**Biology department University of London
Royal Holloway and Bedford New College**

Edward Arnold
A division of Hodder & Stoughton
LONDON   MELBOURNE   AUCKLAND

First published in Great Britain 1991

British Library Cataloguing in Publication Data
Lippold, O.
    Physiology illustrated.
    I. Title   II. Cogdell, B.
    612.0022

    ISBN 0–7131–4531–5

Designed, Illustrated and Typeset in 9pt
Helvetica by Aaron Timbrell and Jonathan
Lippold of Aurora Design, Walton on Thames.
Printed and Bound in Great Britain for Edward
Arnold, a division of Hodder and Stoughton
Limited, Mill Road, Dunton Green, Sevenoaks,
Kent TN13 2YA by Thomson Litho Ltd,
East Kilbride, Scotland.

PREFACE

Learning is a complex process and involves all the sensory modalities. Learning a scientific subject such as Physiology is usually achieved by listening to the spoken word, as in lectures and tutorials, by taking part in experiments, and by various exercises in the techniques of memory recall, for example, essay writing, multiple choice questions and so forth.

There can be little doubt that different students utilise different methods in their study of a scientific discipline, some preferring to rely mainly upon an auditory input, while others concentrate on the written word. These differences reflect the fact that we are individuals and we find that a particular way of learning something suits us better than it suits other people.

There are plenty of textbooks of Physiology that cater for students who find that they can learn easily from the written word; these textbooks moreover are suitable for a wide range of levels from the elementary right up to final year degree standard. Most of them have the major part of their information content in the form of written text with accompanying illustrations. They are not really suitable for those who find that they can learn best from visual material.

As university teachers of Physiology, the authors of this atlas have come to the conclusion that an elementary text, based almost exclusively upon diagrams, graphs and other visual material would be very useful for those students who have visually-orientated memory processes.

An elementary treatment of the subject matter lends itself admirably to visual display especially if the material itself is carefully chosen. We have attempted to deal with the main principles of the science of Physiology at a level that would be expected of any first year Science, Medical, Dental or Nursing student. In addition we envisage the book being of value to students of the ancillary medical subjects such as Psychology, Speech Therapy, Physiotherapy, Radiography, Zoology and even as a reference work for A-level studies. Although the book may seem smaller than the average textbook, this is the result of our diagrammatic method of presentation and is not because anything has been omitted. Our aim is to make the book self-contained and as suitable for those who are studying Physiology for the first time as it is for those who are using it for purposes of revision. In addition, with the advent of the new core curriculum in our schools, it becomes impossible to study Physics, Chemistry and Biology together. A choice must be made. This will result in many students coming to the first year of university or medical school to study Physiology (science or medicine) without a general biological background. This book has been pitched at just the right level for a first year introduction to such a student.

November 1990

Barbara Cogdell
Olof Lippold

ACKNOWLEDGEMENTS

We are very grateful for the considerable help that we have had in preparing this book. Most of the chapters have been tried out during the past four years by our first year science students; their comments and suggestions have been very useful. Dr Lynn Bindman made a detailed criticism of the book for us and it is a pleasure to record our gratitude for all the work that she did. We also thank an anonymous reviewer who took considerable trouble to improve the text and illustrations. The editorial staff at Edward Arnold were invariably helpful and efficient in publishing the book.

Contents

INTRODUCTION

WHAT IS PHYSIOLOGY?

There are many differences between the living and the dead. The nature of these is the province of Physiology; for example how the living body is able to see, hear, keep warm, digest food and so on. A person walks gracefully, lifts an arm accurately and as Physiologists we wonder what is happening in his nervous system, his muscles and his joints. If we want to investigate things like this, we must take the body to pieces in the same sort of way that we might have to examine the engine, or transmission, of a car to see how that works. Obviously one can't study a man or woman by this kind of dismemberment, so tissues and organs such as muscle fibres, or kidneys, have been used after death. Usually we find that the functioning and structures of organs in various animal species are remarkably similar, so it is a reasonable assumption that these are roughly the same in the human species also. We can take these studies a step further by looking at organs and tissues in anaesthetised animals.

Physiology covers the physico-chemical processes taking place in the cells and tissues of the body; the electrical events underlying the actions of the nervous system; and the feedback mechanisms controlling these things, right up to the complex performances of the animal, or human, as a whole. It is only, however, concerned with the normal. Disease mechanisms are the province of Medicine and Pathology, although Clinical Physiology encompasses the study of physiological responses (or compensations) that occur in normal systems when other parts of the body are diseased. For example, the study of changes in the lung or kidneys when the heart goes into failure would be Clinical Physiology. With the rapid expansion of our knowledge, various disciplines have split off from the parent one of Physiology. Biochemistry, the study of chemical processes in cells and tissues was probably the first to split, then Biophysics, dealing with physical processes in cells, and now Neurophysiology is a separate discipline. Quite often physiological research is carried out in departments of Medicine where basic scientific knowl-edge is required for the understanding of disease. The reverse situation is also common; basic research in Physiology reveals in detail how a disease occurs. An often quoted example is how basic research into the pancreas and the hormone insulin led to the discovery of a successful treatment for the then fatal disease of diabetes mellitus. Finally, Physiologists have long been interested in the science underlying such things as athletic perform-ance, the respiratory and nervous systems in an-aesthesia, the effects of low or high barometric pressures, oxygen lack and so on. This is Applied Physiology.

In all these aspects of Physiology it is important to appreciate the role of mechanisms that control bodily functions. We cannot divorce the study of muscular contraction from the feed-back systems that control muscle. In all muscles are structures that signal back to the central nervous system muscle length, and the degree of contractile force. At the spinal level, reflex actions occur to control the performance as precisely as possible in terms of length and force. Without these sophisticated control systems it becomes very difficult to use our muscles properly. Similar mechanisms exist to control arterial blood pressure. Obviously if blood pressure is too high there will be a likelihood of rupture of blood vessels and consequent haemor-rhage; an example being one form of stroke. If blood pressure is too low, the blood supply to the brain will be impaired and consciousness may be lost. In fact, all bodily, physiological processes have control mechanisms to look after their correct, or most advantageous, operation.

Complex feed-back mechanisms control the com-position and temperature of the internal environ-ment, a process that is known as homeostasis. Indeed one could view of the study of Physiology as an exercise in the understanding of homeostasis. The processes of digestion, kidney function, res-piration etc., are all directed towards maintaining the constancy of the internal environment.

The most obvious difference then, between the living and the dead, is that all these control proc-esses are working properly, and in turn the various bodily functions controlled by them are intact. Physiology, on this basis, is the discipline that deals with these bodily functions and their control.

THE CELL

All living organisms (except viruses) are composed of cells. Cells are surrounded by a barrier, the plasma membrane, which separates the interior of the cell, or cytoplasm from the external environment.

• Epithelium which forms the covering or lining of all body surfaces (both external and internal). It includes glands.

• Connective tissue which connects and holds

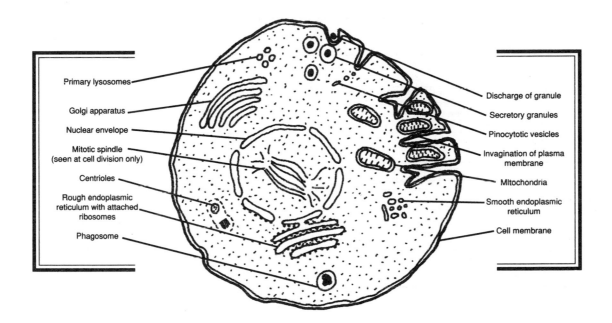

Primary lysosomes

Golgi apparatus

Nuclear envelope

Mitotic spindle
(seen at cell division only)

Centrioles

Rough endoplasmic
reticulum with attached
ribosomes

Phagosome

Discharge of granule

Secretory granules

Pinocytotic vesicles

Invagination of plasma
membrane

Mitochondria

Smooth endoplasmic
reticulum

Cell membrane

A eukaryotic cell contains a nucleus (occasionally more than one), and this is the type of cell that is found in animals, plants, fungi and protista. Eukaryotic cells also have a variety of other membrane-bound cytoplasmic organelles.

Cells come in a variety of shapes and sizes, and few, if any, contain all the components outlined in the table. In the body "like" differentiated cells can combine together to form tissues. There are four basic types of tissue:-

other tissues together. It includes blood, lymph, bones and cartilage.

• Muscle

• Nerve

These basic tissue types are bound together in various combinations to form structural and functional units, the organs. Interacting organs may in turn be grouped together into systems, such as the circulatory system.

Components found in animal cells

STRUCTURE	SHAPE AND SIZE	LOCATION	FUNCTION
1.PLASMA MEMBRANE	Continuous lipoprotein layer 5-10nm thick	Outer surface of all cells	Protects; prevents entry of harmful substances; holds cell together; transport of metabolites; cell to cell recognition and communication; involved in immune responses, hormonal action and conduction of action potentials
2. CYTOPLASMIC ORGANELLES			
Endoplasmic reticulum (ER)	Tubular membrane system; rough ER has attached ribosomes, smooth ER has none	Cytoplasm	Rough ER: protein synthesis Smooth ER: steroid and lipid synthesis
Ribosomes	Spherical 20-35nm across, made of proteins and RNA	Freely suspended or attached to ER	Terminal stages of protein synthesis
Golgi complex or apparatus	Stack of flattened membrane sacs	Cytoplasm	Stores, modifies and packages secretions; manufactures polysaccharides
Secretory vesicles (synaptic vesicles)	Membrane bound vesicles	Cytoplasm	Storage of material to be secreted
Mitochondria	Often sausge-shaped 2μm but can be filamentous or granular, surrounded by two membranes	Few to 100,000 in cytoplasm, more in cells that are active	'Powerhouse' of the cell, terminal stages of aerobic respiration,synthesises ATP
Phagosomes	Membrane bound vesicles	Cytoplasm	Uptake of particulate matter eg large molecules or cells
Pinocytotic vesicles	As above but smaller, less than 0.5μm across	Cytoplasm	Uptake of dissolved materials
Lysosomes	Membrane vesicles about 0.5μm across	Cytoplasm	"Disposal" units, contain hydrolytic enzymes which digest materials; intracellular breakdown (autolysis)
Centrioles	2 cylinders each containing 9 groups of 3 microtubules arranged in a ring, 0.2μm dia 0.4μm long	Lie at right angles to each other, near nuclear membrane	During cell division they separate and move to form the poles of the nuclear spindle
Microtubules	Tubes 15-20nm across, made of protein tubulin	Cytoplasm	Movement and support eg beating of sperm tail or cilia; movement of chromosomes in cell division
Microfilaments	About 5nm across made of polymers of actin and myosin	Cytoplasm	Movement and support eg general flowing motion of cytoplasm (cytoplasmic streaming); contraction of muscle fibres, microvilli
3.NUCLEUS			
Nuclear membrane	Double membrane with pores 70nm across	Surrounds nucleus	Pores serve as transport channels
Nucleoli	Spherical and dense, not enclosed by membrane	1-4 suspended in chromatin of nucleus	Site of synthesis of RNA and DNA
Chromosomal material (chromatin)	Dispersed as fine DNA/protein strands 10-20nm in diameter; during cell division coil to form chromosomes	Nucleus	The information required to synthesise proteins is coded in the DNA.The DNA is copied into "messenger" RNA molecules. These pass into the cytoplasm where the protein synthesis takes place

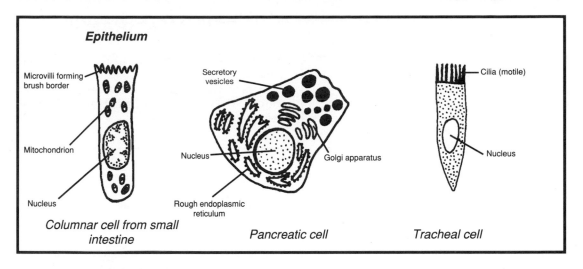

Epithelium

Microvilli forming brush border

Mitochondrion

Nucleus

Columnar cell from small intestine

Secretory vesicles

Nucleus

Rough endoplasmic reticulum

Golgi apparatus

Pancreatic cell

Cilia (motile)

Nucleus

Tracheal cell

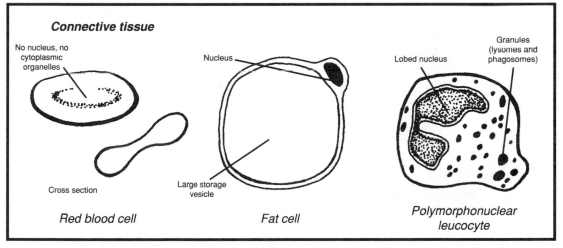

Connective tissue

No nucleus, no cytoplasmic organelles

Cross section

Red blood cell

Nucleus

Large storage vesicle

Fat cell

Lobed nucleus

Granules (lysomes and phagosomes)

Polymorphonuclear leucocyte

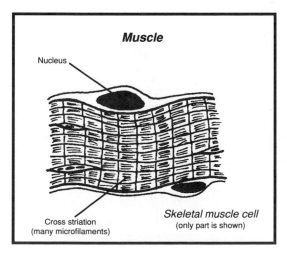

Muscle

Nucleus

Cross striation (many microfilaments)

Skeletal muscle cell
(only part is shown)

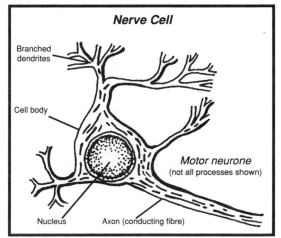

Nerve Cell

Branched dendrites

Cell body

Nucleus

Axon (conducting fibre)

Motor neurone
(not all processes shown)

CELL MEMBRANES

Membranes are of fundamental importance to the structure of cells. They are selectively permeable barriers and act to compartmentalise the interior of the cell. The surface area of membranes within the cell is large.

Estimates of membrane areas in a rat liver cell

Volume μm^3

Cell cytoplasm	5,000
Mitochondria (total)	1,000
Lysosomes (total)	10

Membrane area μm^2

Smooth ER	17,000
Rough ER	30,000
Mitochondrial (inner)	40,000
Mitochondrial (outer)	7,500

In electron micrographs membranes appear as relatively uniform structures approximately 5-10nm thick.

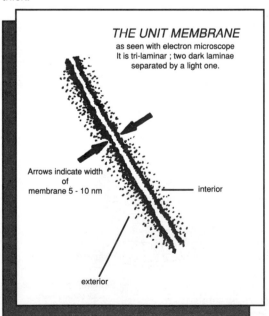

THE UNIT MEMBRANE

as seen with electron microscope
It is tri-laminar ; two dark laminae separated by a light one.

Arrows indicate width of membrane 5 - 10 nm

interior

exterior

The techniques of cell biology

Intensive work by investigators using light microscopes in the mid-ninteenth century, led to the general acceptance of the idea that living things are composed of cells, and that new cells are derived only from preexisting cells (the cell theory). However the resolving power of the light microscope is limited and objects separated by less than about 0.2µm cannot be distinguished. Consequently the appearance of a cell under a light microscope is as shown.

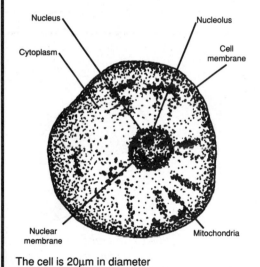

Nucleus

Nucleolus

Cytoplasm

Cell membrane

Nuclear membrane

Mitochondria

The cell is 20µm in diameter
Bacterium 1µm ──────
Large virus 0.1µm ──────── (Invisible)
Small virus 0.01µm ────

Our present detailed knowledge of subcellular organisation has been made possible by electron microscopy and biochemical techniques. Magnification of up to 200,000 times can be obtained with the electron microscope, and objects as small as 1nm can be visualised.

Membranes contain two major classes of compounds, lipids and proteins. In general the more metabolically active a membrane is, the higher the protein/lipid ratio is.

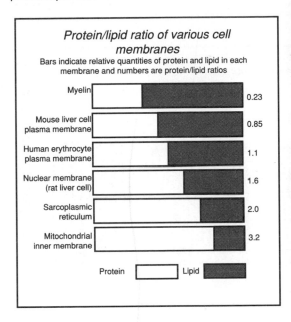

Protein/lipid ratio of various cell membranes

Bars indicate relative quantities of protein and lipid in each membrane and numbers are protein/lipid ratios

Myelin	0.23
Mouse liver cell plasma membrane	0.85
Human erythrocyte plasma membrane	1.1
Nuclear membrane (rat liver cell)	1.6
Sarcoplasmic reticulum	2.0
Mitochondrial inner membrane	3.2

Protein ☐ Lipid ▓

Carbohydrates are also found in membranes linked to both lipids and proteins. The greatest amount of carbohydrate is on the external surface of the plasma membrane.

Membrane lipids

The basic structure of a membrane is provided by its lipid components, which include phospholipids, glycolipids, sphingolipids and sterols. The type and amount of each lipid present is highly variable and depends upon which cellular membrane and which type of cell is being considered.

Lipids are amphipathic molecules i.e. they contain both hydrophilic and hydrophobic regions.

Most membrane lipids have a natural tendency to form bilayers when placed in an aqueous salt solution. The molecules tend to arrange themselves so that the hydrophilic, polar headgroups are in contact with water molecules, while the hydrophobic tails orientate themselves towards other tails.

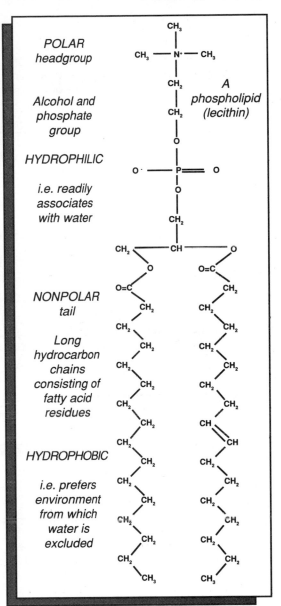

POLAR headgroup

Alcohol and phosphate group

HYDROPHILIC

i.e. readily associates with water

A phospholipid (lecithin)

NONPOLAR tail

Long hydrocarbon chains consisting of fatty acid residues

HYDROPHOBIC

i.e. prefers environment from which water is excluded

Membrane structure

Consideration of the structure of a pure lipid bilayer, together with the observation that membrane proteins are essentially globular in shape, has led to the current fluid-mosaic model for membrane structure (after Singer and Nicolson). The lipid bilayer is thought to provide the framework for the membrane, while the proteins are considered to float individually, or in groups, like icebergs in the sea. Both lipids and proteins are arranged asymmetrically between the bilayer halves of a membrane.

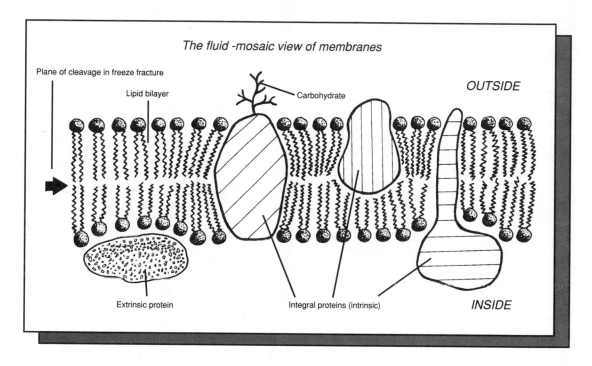

The fluid -mosaic view of membranes

Plane of cleavage in freeze fracture

Lipid bilayer

Carbohydrate

OUTSIDE

Extrinsic protein

Integral proteins (intrinsic)

INSIDE

Evidence for the fluid-mosaic model comes from freeze-fracture electron microscopy studies. The freeze-fracture process exposes the hydrophobic interior of the membrane. In this technique tissues are quick frozen and fractured by a knife edge under vacuum. The line of fracture is down the centre of the membrane bilayer (see previous diagram). After further processing the results may be examined in the electron microscope. Globular particles, 6-18nm across, can be seen penetrating the membrane structure. These are the membrane proteins.

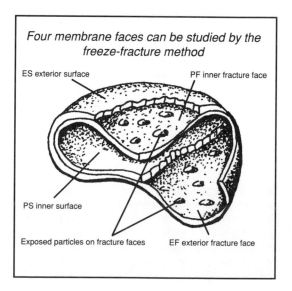

Four membrane faces can be studied by the freeze-fracture method

ES exterior surface

PF inner fracture face

PS inner surface

Exposed particles on fracture faces

EF exterior fracture face

Membrane proteins

Many of the proteins associated with membranes are enzymes. Enzymes are proteins which act as biological catalysts.

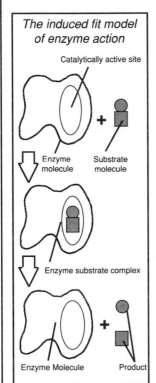

The induced fit model of enzyme action

Catalytically active site

Enzyme molecule + Substrate molecule

Enzyme substrate complex

Enzyme Molecule + Product

The integration of enzymes into membranes provides a means of spatially separating the enzymes, and so preventing undesirable interactions between them. Alternatively, separate enzymes mediating the various steps in a complex series of reactions, can be located together, so that these chemical conversions can proceed with maximum efficiency.

The dynamic nature of membranes

Membranes must not be regarded as rigid static structures. The membrane lipids are in a mobile fluid state and so there is considerable movement of the lipids and proteins within the plane of the membrane. In addition even in tissues with very little growth or cell division , there is a steady turnover of membrane components.

Turnover of cell membrane components	
Component	Half-life i.e. the time required for half substance to be replaced
Phospholipids	15-80 hours
Proteins	50-100 hours

Membranes also change in shape and area. In exocytosis, for example, the membranes of secretory vesicles are added to the plasma membrane. The reverse process, endocytosis, removes sections of the plasma membrane to make closed vesicles within the cytoplasm.

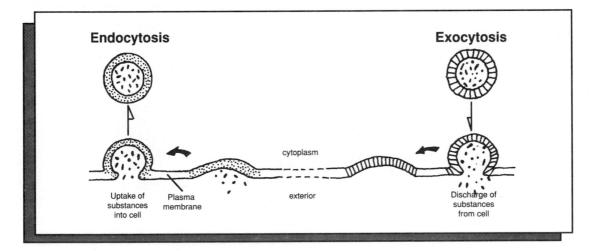

Endocytosis

Exocytosis

cytoplasm

exterior

Uptake of substances into cell

Plasma membrane

Discharge of substances from cell

CELL JUNCTIONS

In tissues the plasma membranes of neighbouring cells are usually 20-30nm apart, but at specialised regions (the intercellular junctions) direct contact is made between them.

• Adhesive junctions or desmosomes. These hold cells together and keep them in fixed positions in tissues. They are disc shaped with diameters of 0.2-0.5μm. They are found in tissues such as the skin which suffer mechanical stress.

• Sealing or tight junctions. These close the space between cells forming a barrier to the diffusion of molecules. Hence two compartments can be physically separated. They extend across the whole interface between one cell and the next. They are found in the blood-brain barrier and at apical regions of epithelia.

• Communicating or gap junctions. These provide channels that allow the flow of ions and other small molecules between cells. They cover areas of 0.1-10μm. They are found in smooth and cardiac muscle.

ENERGY METABOLISM OF CELLS

The energy for the work done by the cell is mainly provided by the mitochondria. A mitochdrion is bounded by two continuous membranes, which enclose two spaces, an intermembrane space and an inner matrix space. The energy yielding reactions take place in and on the inner mitochondrial membrane.

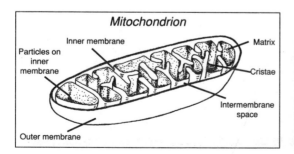

Mitochondrion

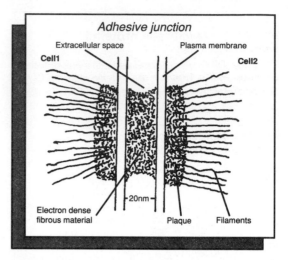

Adhesive junction

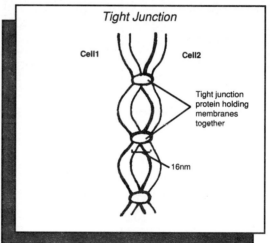

Tight Junction

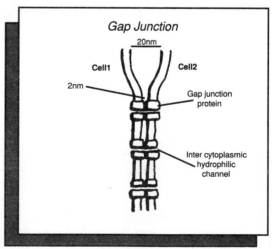

Gap Junction

Metabolic fuels produced from the breakdown of carbohydrates, proteins and fats, are delivered to the mitochondrion. These compounds are then oxidised by the process of oxidative phosphorylation and energy is released. Some of this energy is conserved as adenosine triphosphate molecules (ATP). ATP serves as a universal "energy currency" that can be spent in powering other functions of cells, such as synthesis of more complex compounds, muscular contraction and nerve conduction.

SECRETION

The process of secretion begins on the membranes of the rough endoplastic reticulum (ER). Here protein is synthesised by the ribosomes. The ribosomes decode a molecule of messenger RNA and translate it into a sequence of amino acid residues, which are linked together to form the desired protein.

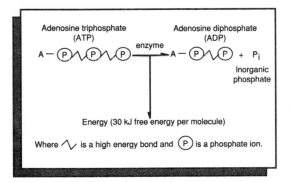

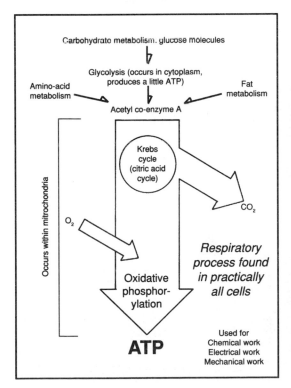

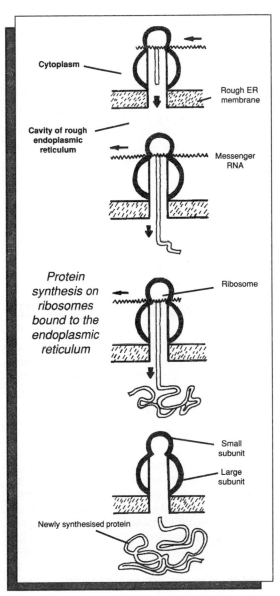

From their site of synthesis, the secretory proteins are transported into the Golgi apparatus. Here, after any necessary chemical modifications such as the addition of carbohydrate, they are packaged into secretory vesicles. Finally the secretory prod-

useful material may be absorbed into the cytoplasm, or alternatively the contents of the vesicle may be released back through the plasma membrane by exocytosis.

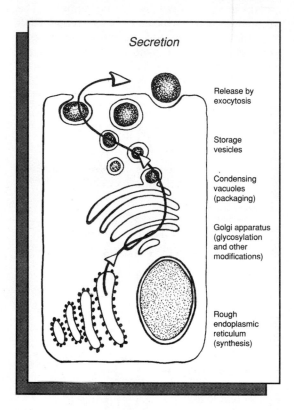

Secretion

Release by exocytosis

Storage vesicles

Condensing vacuoles (packaging)

Golgi apparatus (glycosylation and other modifications)

Rough endoplasmic reticulum (synthesis)

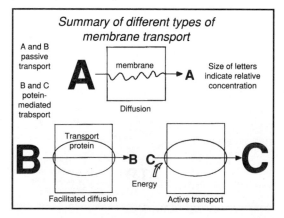

Pinocytosis and Phagocytosis

Pinocytosis

Phagocytosis

Pinocytotic vesicle

Coalescence of phagosome and lysosome

Phagosome

Plasma membrane

Lysosome

Digestion

Absorption of digestive products

Discharge of waste

ucts are either discharged continuously from the cell, or they remain in the vesicles until an appropriate stimulus induces their release to the exterior.

PHAGOCYTOSIS AND PINOCYTOSIS

Phagocytosis refers to the engulfing of particulate matter (such as the engulfing of bacteria by larger cells), while pinocytosis is the uptake of particulate-free extracellular fluid. These two processes are accomplished by endocytosis (see earlier). The pinocytotic vesicles and the phagosomes may then coalesce with lysosomes. This brings the digestive enzymes of the lysosomes into contact with the contents of the vesicles. As the digestion proceeds

MEMBRANE TRANSPORT

Molecules may move across membranes either passively or actively.

Summary of different types of membrane transport

A and B passive transport

B and C potein-mediated trabsport

A — membrane — A

Size of letters indicate relative concentration

Diffusion

B — Transport protein — B   C — Energy — C

Facilitated diffusion

Active transport

Passive transport

Passive transport of a substance across a membrane relies on the process of diffusion and is a spontaneous, non-energy requiring event. Diffusion is the process whereby solutes will gradually spread from a region of high concentration to one of low concentration, until they are evenly distributed throughout the solution.

The same process of diffusion will take place between two solutions of differing concentrations when they are separated by a permeable membrane. The diffusion, or net movement of solute across the membrane, will continue until the molecules of solute are evenly distributed on both sides.

Diffusion of a substance through a membrane into a cell or compartment will therefore occur if the following conditions are met:-

   • There is a higher concentration of the substance outside the cell (i.e. the substance will be moving down a concentration gradient).

   • The membrane is permeable to the substance.

Small, lipid soluble molecules are freely permeable through biological membranes, while other more hydrophilic molecules such as urea, methanol and certain ions, have a much more limited permeability. A range of larger hydrophilic molecules also cross membranes by diffusion, but in this case their diffusion is facilitated by specific, membrane-bound carrier proteins. Again this process is also non-energy requiring, and always transports substances from the compartment of higher concentration to that of lower concentration. The transport of glucose out of the capillaries in the brain (i.e. across the blood-brain barrier) is an example of facilitated diffusion.

Osmosis

Water is freely permeable across most biological membranes. The simple process of diffusion is complicated by the fact that biological membranes are semipermeable. A semipermeable membrane will allow some of the molecules or ions to pass

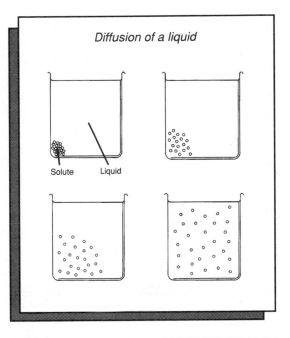
Diffusion of a liquid

Solute   Liquid

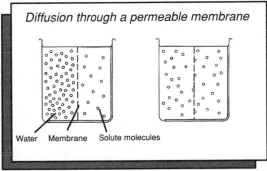
Diffusion through a permeable membrane

Water   Membrane   Solute molecules

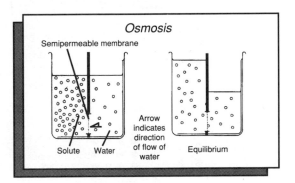
Osmosis

Semipermeable membrane

Solute   Water   Arrow indicates direction of flow of water   Equilibrium

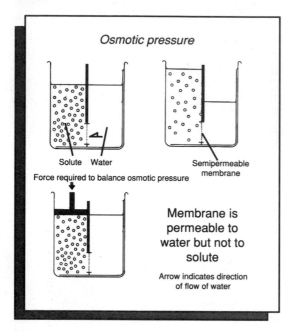

Osmotic pressure

Solute   Water

Force required to balance osmotic pressure

Semipermeable membrane

Membrane is permeable to water but not to solute

Arrow indicates direction of flow of water

through much more readily than others. For example when two solutions with differing concentrations are separated by a semipermeable membrane (which allows water to pass through but not the solute molecules), the diffusion of water in one direction will exceed that in the other. Water will move across the membrane from the solution with the lower solute concentration to that with the higher concentration, until the concentrations of the two solutions are equalised. This net flow of water is called osmosis.

The osmotic flow of water across a membrane may be prevented by applying an opposing hydrostatic pressure. The pressure that is just sufficient to prevent the movement is called the osmotic pressure. The osmotic pressure of a solution is dependent only on the number of dissolved particles (undissociated molecules or ions) and not the kind of molecules present.

There are a variety of mechanisms (in particular those of the kidney) which ensure that the osmotic pressure of the internal environment is maintained at a constant level. The cells of the body do not possess their own osmotic regulatory mechanisms and will not survive if placed in hypotonic or hypertonic solutions. For example if a red blood cell is placed in distilled water it will swell up and burst, releasing its haemoglobin. This is known as haemolysis.

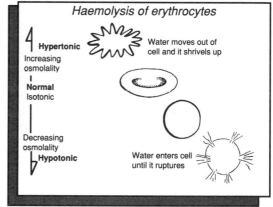

Haemolysis of erythrocytes

Hypertonic
Increasing osmolality

Normal
Isotonic

Decreasing osmolality
Hypotonic

Water moves out of cell and it shrivels up

Water enters cell until it ruptures

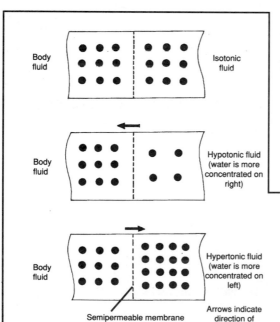

Body fluid | Isotonic fluid

Body fluid | Hypotonic fluid (water is more concentrated on right)

Body fluid | Hypertonic fluid (water is more concentrated on left)

Semipermeable membrane

Arrows indicate direction of water flow

The concentration of particles in a solution can be expressed as osmolality. Isotonic solutions have the same effective osmolality as body fluids (about 285 milliosmolals).

Hypotonic fluids have a lower osmolality than body fluids.

Hypertonic fluids have an effective osmolality greater than that of body fluids.

Diffusion of ions

When an uncharged molecule moves across a membrane just the chemical species moves. However when a charged molecule moves, there is also the transfer of an electrical charge across the membrane. Consequently the movement of charged molecules depends not only on concentration gradients, but also on electrical gradients or membrane potentials.

Biological membranes are not equally permeable to all ions. In particular, cells contain large, negatively charged protein molecules which cannot cross the plasma membrane. The balance point between the effects of the concentration gradient of the small, permeable ions and the large, charged, impermeable proteins and their electrical potential gradient, is reached when the Nernst equation is satisfied.

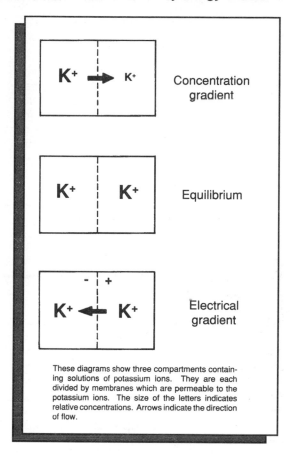

These diagrams show three compartments containing solutions of potassium ions. They are each divided by membranes which are permeable to the potassium ions. The size of the letters indicates relative concentrations. Arrows indicate the direction of flow.

Nernst equation

$$E = \frac{RT}{nF} \; \ln \; \frac{C_1}{C_2}$$

where n is the net charge of the ion

o.g. n=1 for K^+ and Cl
n=2 for Mg^{++} and SO4^{--}

E is the electrical potential difference across the membrane

F is the Faraday constant

R is the gas constant

T is the absolute temperature in deg C

C$_1$ and **C$_2$** are the concentrations of the ion on either side of the membrane

Active transport

Active transport moves molecules or ions across membranes against gradients of concentration or electrical potential. Energy, usually in the form of ATP, is required for this "uphill" movement. The transport is catalysed by membrane-bound transport proteins (permeases). Permeases, like enzymes, exhibit the property of specificity, and only accept those molecules which "fit" into their binding sites.

Active transport systems are important in absorption of nutrients from the intestinal contents; the recovery of sugars and amino acids from the glomerular filtrate in the kidney, and the creation and maintenance of ionic gradients.

The sodium pump

The active transport of sodium and potassium ions is bought about by an enzyme, Na$^+$/K$^+$ATPase, which is located in the plasma membrane. Na$^+$/K$^+$ATPase hydrolyzes ATP at the cytoplasmic surface of the plasma membrane. On average for every ATP molecule consumed, three sodium ions are transported outwards and two potassium ions pumped inwards. This is an electrogenic pump (i.e. it generates a potential).

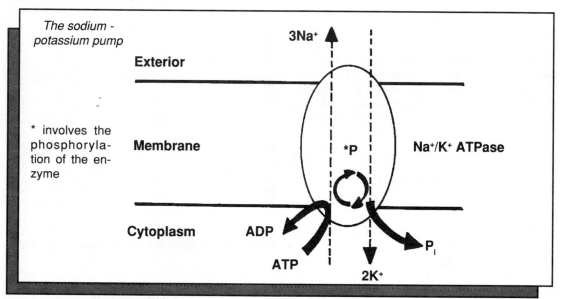

The sodium - potassium pump

* involves the phosphorylation of the enzyme

The sodium-potassium pump works to maintain the sodium and potassium ion concentration gradients characteristic of vertebrate cells i.e. the concentration of potassium ions is higher inside the cell than outside, while the opposite is true for sodium ions. The functions of the sodium-potassium pump activity includes the following:

• Producing the membrane potential across nerve cells (the resting potential).

• Establishing sodium ion gradients that drive the active transport of sugars and amino acids into cells.

• Causing the excretion of sodium ions that helps to reduce the osmotic pressure within cells.

BODY FLUIDS AND BLOOD

BODY FLUIDS

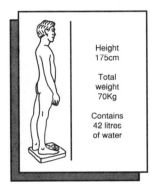

Height
175cm

Total
weight
70Kg

Contains
42 litres
of water

Sixty percent of the weight of an average man is water. This percentage will vary depending on such factors as the degree of fatness, age and sex.

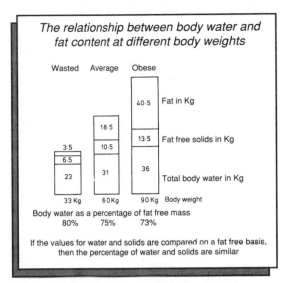

The relationship between body water and fat content at different body weights

	Wasted	Average	Obese	
			40·5	Fat in Kg
		18·5		
	3·5	10·5	13·5	Fat free solids in Kg
	6·5			
	23	31	36	Total body water in Kg
Body weight	33 Kg	60 Kg	90 Kg	

Body water as a percentage of fat free mass
80%   75%   73%

If the values for water and solids are compared on a fat free basis,
then the percentage of water and solids are similar

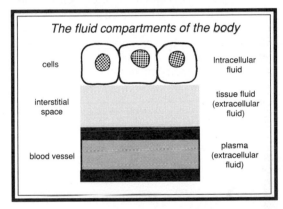

The fluid compartments of the body

cells — Intracellular fluid

interstitial space — tissue fluid (extracellular fluid)

blood vessel — plasma (extracellular fluid)

there is a clear separation between the plasma and tissue fluid. The blood plasma does not come into direct contact with the body's cells apart from those which form the walls of the blood vessels and the blood cells themselves.

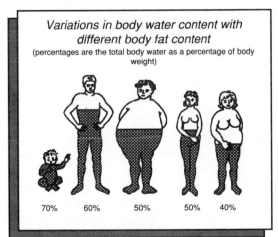

Variations in body water content with different body fat content
(percentages are the total body water as a percentage of body weight)

70%   60%   50%   50%   40%

The most significant of these factors is body fat content. Adipose tissue or fat has a much lower water content (between 10 and 30%) than other tissues except compact bone. Most tissues have very similar water contents (in excess of 70%).
The body is divided into two distinct fluid compartments, the intracellular fluid and the extracellular fluid. Extracellular fluid is further divided into blood plasma and tissue or interstitial fluid. In mammals

Relative amounts of fluid in the body fluid compartments

Percentages are the amount of fluid as a percentage of body weight
Total body water = 42 litres (60% of total body weight)

40% 28 litres	15% 10.5 litres Tissue fluid	5% 3.5 litres Plasma
Intracellular Fluid 40% (2/3 of total body water)	Extracellular Fluid 20% (1/3 total body water)	

Intracellular and extracellular fluid vary considerably in their compositions, but tissue fluid and plasma are almost identical except plasma contains more protein.

Concentration of various components in body fluids (mmol/litre)			
Substance	plasma	interstitial fluid	intracellular fluid
sodium ions	140	145	10
potassium ions	4	4	160
chloride ions	100	115	3
bicarbonate ions	28	30	10
protein	16	10	55

The regulation of both the volume and composition of body fluids involves the control of water intake and loss by a variety of means. It is important in this context that although the intracellular fluid, tissue fluid and plasma are separated into distinct compartments, there is a continuous movement of water and other substances between them.

BLOOD

Blood is the body's major transport system. It has the following important functions:-

• It carries respiratory gases, nutrients and wastes.

• It distributes heat.

• It passes through monitoring organs and enables the detection of variations in hormone levels, osmotic pressure, pH, temperature etc.

• It provides cellular transport.

• It contains agents which protect against infections and tumours.

Blood flows around the body in well defined vessels which form a closed circulatory system separate from the general tissue fluids.

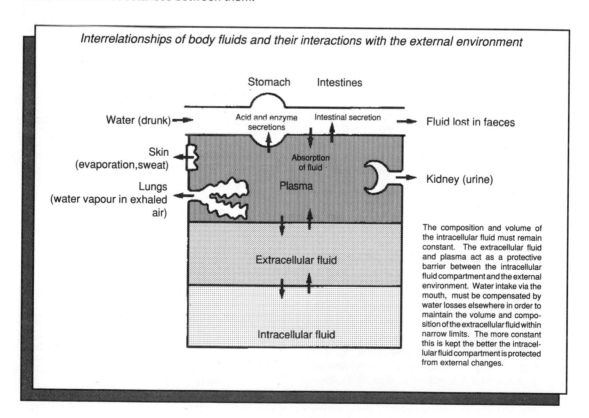

Interrelationships of body fluids and their interactions with the external environment

Stomach   Intestines

Water (drunk) →   Acid and enzyme secretions   Intestinal secretion   → Fluid lost in faeces

Skin (evaporation, sweat)

Absorption of fluid

→ Kidney (urine)

Lungs (water vapour in exhaled air)

Plasma

Extracellular fluid

Intracellular fluid

The composition and volume of the intracellular fluid must remain constant. The extracellular fluid and plasma act as a protective barrier between the intracellular fluid compartment and the external environment. Water intake via the mouth, must be compensated by water losses elsewhere in order to maintain the volume and composition of the extracellular fluid within narrow limits. The more constant this is kept the better the intracellular fluid compartment is protected from external changes.

Composition of blood

Blood is classified as connective tissue and is a thick suspension of cells in a watery electrolyte solution. When whole blood is centrifuged in a test tube, it separates into plasma and cells (provided clotting is prevented).

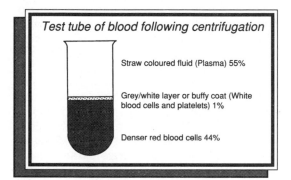

Test tube of blood following centrifugation

Straw coloured fluid (Plasma) 55%

Grey/white layer or buffy coat (White blood cells and platelets) 1%

Denser red blood cells 44%

The percentage of red cells in the blood by volume is known as the haematocrit, and the normal range of values for this is 40-47%. If the blood is left to stand, it clots into a jellylike mass. The fluid that remains after clotting is called serum. Serum is plasma minus the blood clotting protein, fibrinogen.

PLASMA

Plasma is a clear liquid, which remains after the blood cells are removed. Following a meal it may appear opalescent or milky due to the presence of fatty substances. The constituents of plasma are:-

• water (≈90%)

• protein (≈8%)

• inorganic ions (≈1%), mostly in the form of sodium chloride and sodium bicarbonate

• organic acids and other substances (≈1%) requiring transport around the body

(percentages given are those of whole plasma by weight)

Plasma proteins

The major types of plasma proteins are:-

	Concentration	Molecular weight
Albumins	35 - 45 g/l	~69,000
Globulins	15 - 32 g/l	150,00 - 1,000,000
Fibrinogen	2 - 4 g/l	400,000

The globulins include α- and β-globulins, and immunoglobulins. Almost all plasma proteins are synthesised in the liver, the major exception being the immunoglobulins which are manufactured by cells of the immune system.

The plasma proteins serve a number of general functions.

• They are responsible for exerting the osmotic pressure of blood. Hence they maintain the water balance between plasma and tissue fluid.

• They produce the viscosity of blood. This affects the maintenance of normal blood pressure by the heart.

• They provide about 1/6th of the total buffering power of the blood.

• They transport substances circulating in the blood. Several hormones, fatty acids and other lipids, vitamins and various minerals are carried in combination with proteins.

Some plasma proteins have special functions. Fibrinogen, for example, is essential for blood clotting. The immunoglobulins (antibodies) are important in the body's defence against invading foreign organisms.

Inorganic ions

The principal inorganic ions in the plasma are sodium, bicarbonate and chloride ions. Potassium, calcium and magnesium ions are also present, but in lesser amounts. The concentrations of these ions are controlled precisely, mainly by the kidney.

Organic nutrients

The plasma transports numerous organic nutrients from the gastrointestinal tract and storage areas, such as the liver, to other sites in the body. These nutrients include glucose, fats (phospholipids and cholesterol) and amino acids.

Nitrogenous waste products

Nitrogenous waste products, such as urea, ammonia and uric acid, are transported to the kidneys for excretion.

Hormones

These are important regulatory chemicals which travel in the blood stream from their site of production to their site of action.

Gases

Nitrogen (which is inert), oxygen and carbon dioxide are dissolved in the plasma.

BLOOD CELLS

There are three types of blood cells: red blood cells, platelets and white blood cells.

The red blood cells or erythrocytes are biconcave discs without nuclei. They contain haemoglobin which gives the blood its red colour and transports oxygen.

The platelets or thrombocytes are very small cells without nuclei, that play a role in blood clotting.

The white blood cells or leucocytes are colourless cells, and may be divided into granulocytes, monocytes and lymphocytes. The granulocytes, which are distinguished by the presence of cytoplasmic granules, are also known as polymorphonuclear leucocytes (polymorphs) as their nuclei are divided into several lobes. Granulocytes may be further subdivided according to the staining reactions of their granules into neutrophils, eosinophils and basophils. The leucocytes are concerned in the defence of the body against infection. They are not restricted to the blood vessels and are found in large quantities in the lymphatic system and loose connective tissues.

> The life cycle of blood cells
>
> The formation of blood cells is called haemopoiesis. In the adult haemopoiesis occurs in the bone marrow, in particular in the sternum, ribs, vertebrae, cranium and pelvis. In the fetus haemopoiesis also takes place in the liver and spleen.

The types of blood cells

Type of blood cell	Appearance in a stained blood film (diameters as measured from film)	Number of cells per mm³ of blood	Type of blood cell	Appearance in a stained blood film (diameters as measured from film)	Number of cells per mm³ of blood and percentage of total number of white blood cells	
			White blood cells			
Red blood cells	○ 7µm	$4.8 - 5.4 \times 10^6$	Neutrophils		10 - 14 µm	2,500-7,500 60%
			Eosinophils		10 - 14 µm	40 - 400 3%
			Basophils		10 - 14 µm	15 - 100 <1%
Platelets	2 - 5µm	$150 - 500 \times 10^3$	Monocytes		15 - 25 µm	200 - 800 5%
			Lymphocytes		7 - 14 µm	1,500-3,500 30%

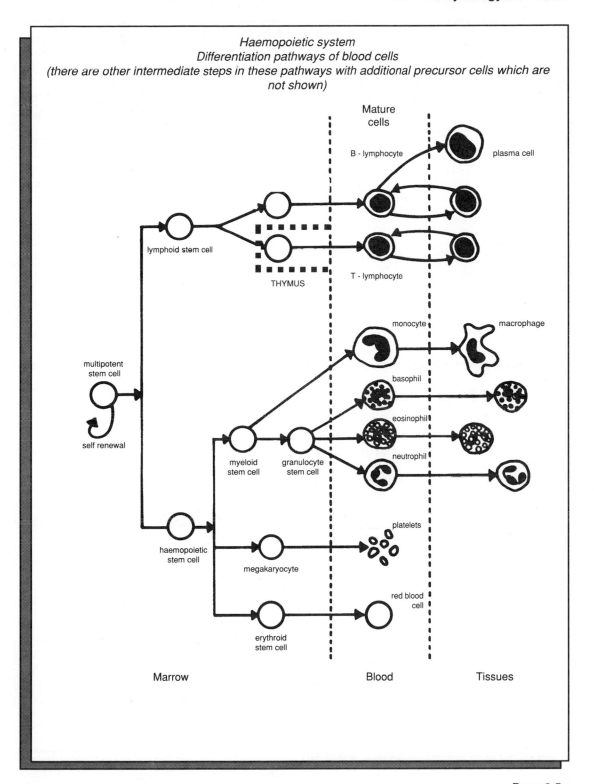

The life history of blood cells

CELL	ORIGIN	LIFESPAN OF MATURE CELL	METHOD OF REMOVAL
Red blood cells	Bone marrow	120 days	Reticulo-endothelial system, components such as iron and amino acids are recycled.
Platelets	Megakaryocytes in bone marrow	8 - 14 days	Reticulo-endothelial system.
White blood cells			
Neutrophils	Bone marrow	Strored in marrow up to 11 days. Only in blood for 3-12 hours as transit. May live 12 days in tissues, but they cannot return to blood stream.	Reticulo-endothelial system, or by leaving the body.
Eosinophils	Bone marrow		
Basophils	Bone marrow		
Monocytes	Bone marrow precursor cells of macrophages, which may also reproduce themselves.	Blood transit time is 32 hours. May live several months in tissues as macrophages.	Reticulo-endothelial system, or by leaving the body.
Lymphocytes	Bone marrrow and other lymphoid tissues such as thymus, spleen and lymph nodes.	Continually recirculate between blood and lymphatic system. Long lived, average 4.5 years with some over 10 years.	Reticulo-endothelial system.

Red blood cells or erythrocytes

The function of red blood cells is to transport the respiratory gases, oxygen and carbon dioxide, around the body. They may be viewed as sacs, bounded by a plasma membrane, containing haemoglobin.

Red blood cells moving along a small blood vessel

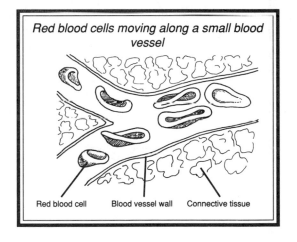

Red blood cell   Blood vessel wall   Connective tissue

The red blood cell or erythrocyte

The centre of the red blood cell is less densely coloured than the periphery because of its bi-concave disc shape.

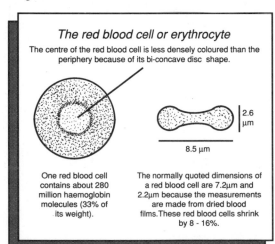

2.6 μm

8.5 μm

One red blood cell contains about 280 million haemoglobin molecules (33% of its weight).

The normally quoted dimensions of a red blood cell are 7.2μm and 2.2μm because the measurements are made from dried blood films.These red blood cells shrink by 8 - 16%.

The biconcavity of the red blood cells is functionally useful, in that it allows a much larger surface area for diffusion for a given enclosed volume than a sphere would. It also provides for diffusion to take place across the shortest possible distance between the membrane and points in the cell interior.

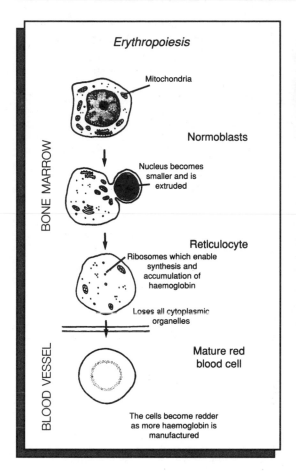

Erythropoiesis

Mitochondria

Normoblasts

Nucleus becomes smaller and is extruded

Reticulocyte

Ribosomes which enable synthesis and accumulation of haemoglobin

Loses all cytoplasmic organelles

Mature red blood cell

The cells become redder as more haemoglobin is manufactured

BONE MARROW

BLOOD VESSEL

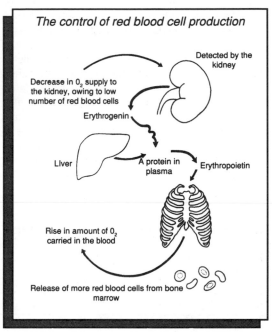

The control of red blood cell production

Detected by the kidney

Decrease in O_2 supply to the kidney, owing to low number of red blood cells

Erythrogenin

Liver

A protein in plasma

Erythropoietin

Rise in amount of O_2 carried in the blood

Release of more red blood cells from bone marrow

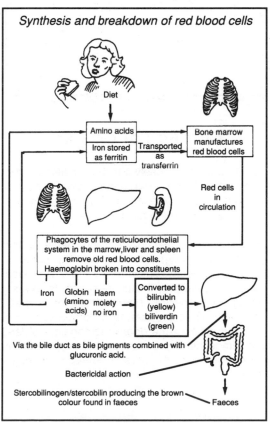

Synthesis and breakdown of red blood cells

Diet

Amino acids

Iron stored as ferritin

Transported as transferrin

Bone marrow manufactures red blood cells

Red cells in circulation

Phagocytes of the reticuloendothelial system in the marrow, liver and spleen remove old red blood cells. Haemoglobin broken into constituents

Iron | Globin (amino acids) | Haem moiety no iron

Converted to bilirubin (yellow) biliverdin (green)

Via the bile duct as bile pigments combined with glucuronic acid.

Bactericidal action

Stercobilinogen/stercobilin producing the brown colour found in faeces

Faeces

Red blood cells are remarkably deformable. They can flex and twist into different shapes, so that they are able to squeeze through the smallest blood vessels.

Old red blood cells are continually being removed from the circulation. So in order to maintain a constant number of circulating red blood cells, about two hundred thousand million new red blood cells need to be made each day (i.e. 2×10^{11} cells/day).

In certain disorders there may be fewer than normal circulating red blood cells, or the red blood cells themselves may be abnormal. This leads to a haemoglobin concentration below that of normal, and the condition known as anaemia.

The reticuloendothelial system removes red blood cells from the blood stream at the end of their lives.

This system consists of phagocytic cells scattered throughout the body and includes macrophages.

The haemoglobin from the old red blood cells is broken down into its constituents which are either recycled or excreted. If there is any blockage or failure of the excretory pathway, the bile pigments, bilirubin and biliverdin, accumulate in the blood and cause the skin to appear yellow. This condition is known as jaundice and is generally indicative of liver or gall bladder disease. These pigments also cause the skin discolouration in bruising.

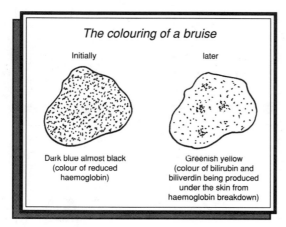

The colouring of a bruise

Initially

later

Dark blue almost black
(colour of reduced
haemoglobin)

Greenish yellow
(colour of bilirubin and
biliverdin being produced
under the skin from
haemoglobin breakdown)

Carriage of oxygen

Oxygen combines loosely with the iron atoms of the haemoglobin to form oxyhaemoglobin. This process is known as oxygenation. During oxygenation the iron remains in the ferrous state.

Oxygenation is a reversible reaction. Hence, as the blood passes through the lungs, haemoglobin combines with oxygen. When this oxygenated blood reaches the tissues, it then releases its oxygen leaving - oxygen-free - haemoglobin (de-oxyhaemoglobin). This process is accompanied by a change in colour; the deoxyhaemoglobin is dark blue, almost black in colour, and on oxygenation it is converted to the bright red coloured oxyhaemoglobin.

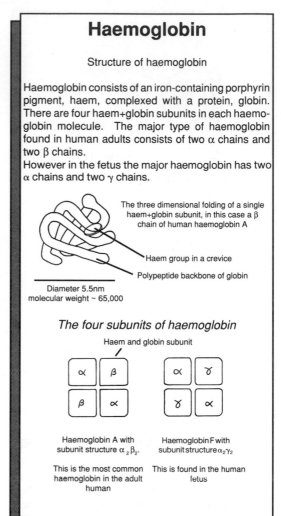

Haemoglobin

Structure of haemoglobin

Haemoglobin consists of an iron-containing porphyrin pigment, haem, complexed with a protein, globin. There are four haem+globin subunits in each haemoglobin molecule. The major type of haemoglobin found in human adults consists of two α chains and two β chains.
However in the fetus the major haemoglobin has two α chains and two γ chains.

The three dimensional folding of a single haem+globin subunit, in this case a β chain of human haemoglobin A

Haem group in a crevice

Polypeptide backbone of globin

Diameter 5.5nm
molecular weight ~ 65,000

The four subunits of haemoglobin

Haem and globin subunit

| α | β |
| β | α |

| α | γ |
| γ | α |

Haemoglobin A with
subunit structure $\alpha_2\beta_2$.

Haemoglobin F with
subunit structure $\alpha_2\gamma_2$

This is the most common
haemoglobin in the adult
human

This is found in the human
fetus

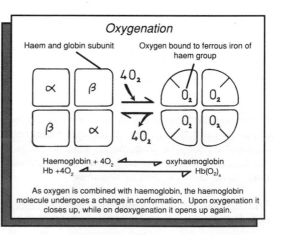

Oxygenation

Haem and globin subunit

Oxygen bound to ferrous iron of haem group

| α | β |
| β | α |

$4O_2$

| O_2 | O_2 |
| O_2 | O_2 |

$4O_2$

Haemoglobin + $4O_2$ ⇌ oxyhaemoglobin
Hb + $4O_2$ ⇌ $Hb(O_2)_4$

As oxygen is combined with haemoglobin, the haemoglobin molecule undergoes a change in conformation. Upon oxygenation it closes up, while on deoxygenation it opens up again.

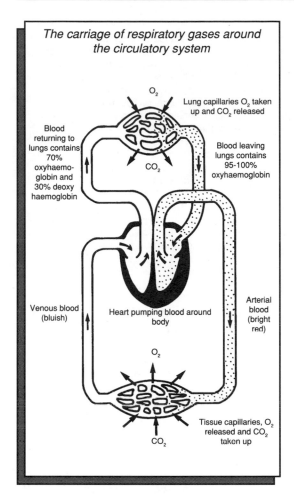

The carriage of respiratory gases around the circulatory system

O_2

Lung capillaries O_2 taken up and CO_2 released

Blood returning to lungs contains 70% oxyhaemoglobin and 30% deoxy haemoglobin

CO_2

Blood leaving lungs contains 95-100% oxyhaemoglobin

Venous blood (bluish)

Heart pumping blood around body

Arterial blood (bright red)

O_2

Tissue capillaries, O_2 released and CO_2 taken up

CO_2

Haemoglobin will also combine readily with other gases. For example, haemoglobin and carbon monoxide combine to form the highly stable carboxyhaemoglobin. Their great affinity accounts for the extremely lethal nature of carbon monoxide inhalation. Carboxyhaemoglobin has a bright cherry red colour which causes the characteristic skin colouring following carbon monoxide poisoning.

Oxygenation should not be confused with oxidation which would occur if the iron changed from the reduced ferrous form into its oxidized ferric form. Oxidation can happen in certain blood diseases. The oxidized form of haemoglobin, methaemoglobin, is incapable of transporting oxygen, and so methaemoglobin formation is harmful.

Oxygen dissociation curves.

The relationship between the oxygen uptake by haemoglobin and the concentration of the oxygen in the surroundings, is described by an oxygen dissociation curve. It is usual to plot the percentage saturation of haemoglobin (Hb) with oxygen against the partial pressure of oxygen (PO_2).

The percentage saturation of Hb with O_2

$$\frac{\text{amount of } O_2 \text{ actually combined with Hb}}{\text{maximum amount of } O_2 \text{ which can combine with Hb}} \times 100$$

If oxygen uptake by the blood is impaired during its passage through the lungs, then even arterial blood will be blue. When a patient's colour becomes blue, this is termed cyanosis.

Normal blood contains 145g haemoglobin in each litre. One gram of haemoglobin can combine with 1.34ml of oxygen. Therefore every litre of blood is able to carry a maximum of 145x1.34=200ml of oxygen. However, if there was no haemoglobin, a litre of blood could only dissolve and transport at most 3ml of oxygen. Consequently the presence of haemoglobin enables the blood to transport 70 times as much oxygen.

The oxygen dissociation curve for haemoglobin is always sigmoidal (S-shaped). This sigmoidal shape arises because the haemoglobin molecule exhibits the property of cooperativity. When a haemoglobin subunit takes up oxygen, it undergoes a conformational change. This alters the structural relationship between this and the other three globin subunits. As a result, the binding of oxygen to the other haem groups is facilitated. Hence at low partial pressures of the oxygen, the dissociation curve rises slowly because the binding of the first oxygen molecule is relatively difficult. The curve will then rise more steeply as the additional binding of oxygen has been facilitated. Finally the curve levels off as the haemoglobin becomes saturated with oxygen.

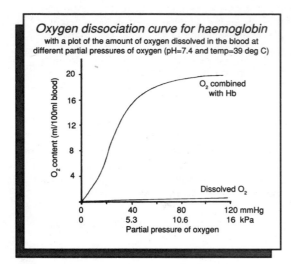

Oxygen dissociation curve for haemoglobin
with a plot of the amount of oxygen dissolved in the blood at
different partial pressures of oxygen (pH=7.4 and temp=39 deg C)

Haemoglobin, because of its S-shaped oxygen dissociation curve, is ideally suited as an oxygen carrier in the blood. The slope of the curve for PO_2 values above 60mmHg is very shallow. This means that most of the haemoglobin (90% or more) leaving the lungs, will be oxygenated even if there is a considerable drop in the oxygen content of the inspired air. The steepest portion of the curve is in the range of the oxygen partial pressures found in tissue fluids. Here small changes in PO_2 such as occur during exercise, will lead to a large release of oxygen to the tissues. For example at rest when the demand for oxygen is low, the PO_2 in the tissues will be 40mmHg and the blood will give up approximately 30% of its oxygen load. However, during exercise the PO_2 in the tissues can drop to 20mmHg. This then will result in a large increase in the amount of oxygen released.

Factors affecting oxygen affinity and shape of dissociation curve

The partial pressure of oxygen at which haemoglobin is half saturated is called the P_{50}. The P_{50} of haemoglobin can vary depending upon a variety of factors. These are:-

- changes in pH

- changes in temperature

- changes in concentration of 2,3-diphosphoglycerate

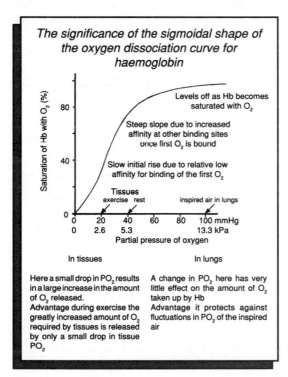

The significance of the sigmoidal shape of the oxygen dissociation curve for haemoglobin

In tissues

Here a small drop in PO_2 results in a large increase in the amount of O_2 released.
Advantage during exercise the greatly increased amount of O_2 required by tissues is released by only a small drop in tissue PO_2

In lungs

A change in PO_2 here has very little effect on the amount of O_2 taken up by Hb
Advantage it protects against fluctuations in PO_2 of the inspired air

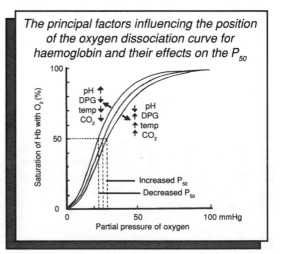

The principal factors influencing the position of the oxygen dissociation curve for haemoglobin and their effects on the P_{50}

Effect of pH (Bohr effect)

A decrease in pH will shift the oxygen dissociation curve to the right and cause haemoglobin to release more oxygen at any given partial pressure of oxygen. This ensures that the haemoglobin releases oxygen readily in the tissues where the pH drops as a result of an increase in the concentration of CO_2. Such an increase in CO_2 typically occurs during periods of muscle activity. The Bohr effect then ensures an increased supply of oxygen during exercise.

Effect of temperature

Increasing temperature reduces the oxygen affinity of haemoglobin, and so shifts the dissociation curve to the right. Usually a rise in temperature in a tissue results from increased metabolic activity. This increased activity requires more oxygen, and the effect of temperature upon the oxygen affinity of haemoglobin satisfies this need.

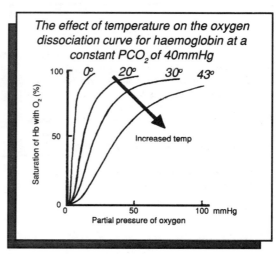

The effect of temperature on the oxygen dissociation curve for haemoglobin at a constant PCO_2 of 40mmHg

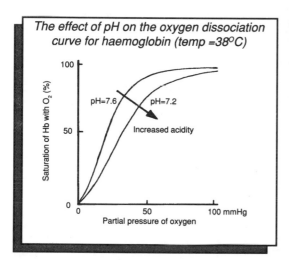

The effect of pH on the oxygen dissociation curve for haemoglobin (temp =38°C)

Concentration of 2,3-diphosphoglycerate (2,3-DPG)

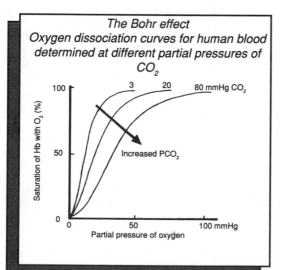

The Bohr effect
Oxygen dissociation curves for human blood determined at different partial pressures of CO_2

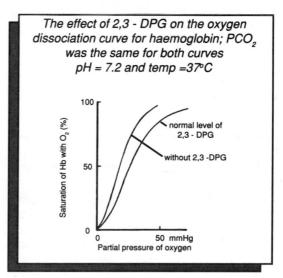

The effect of 2,3 - DPG on the oxygen dissociation curve for haemoglobin; PCO_2 was the same for both curves
pH = 7.2 and temp =37°C

2,3-DPG is produced by a modification of glycolysis which is peculiar to red blood cells. When 2,3-DPG combines with haemoglobin, it alters the haemoglobin's affinity for oxygen. A rise in the concentration of 2,3-DPG reduces the oxygen affinity of haemoglobin. 2,3-DPG is extremely important in several situations. For example, in the transfer of oxygen across the placental barrier and in the adaptation to living at high altitude.

> The exchange of oxygen between the fetus and the mother across the placental barrier occurs because fetal haemoglobin has a much lower affinity for 2,3-DPG. In the absence of 2,3-DPG, human adult and fetal haemoglobin have very similar oxygen dissociation curves. In the maternal placental blood the presence of 2,3-DPG shifts the oxygen dissociation curve of haemoglobin A to the right. This promotes oxygen dissociation. In the fetal blood, the oxygen affinity of haemoglobin F is not changed by 2,3-DPG. Hence haemoglobin F maintains a higher affinity for oxygen than the haemoglobin A, and so oxygen is transferred from the mother to the fetus.
>
> *Oxygen dissociation curves for adult and fetal haemoglobin*
>
>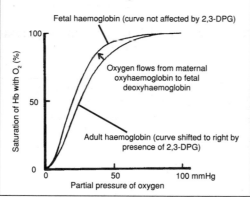

2,3-DPG also plays a role in adaptation to living at high altitude. When you go from sea level to high altitude (4500 metres), the concentration of 2,3-DPG in red blood cells increases from 4.5 to 7.0 mM within two days. This results in an extra release of oxygen from the haemoglobin to the tissues and so compensates for the decreased PO_2 at high altitudes.

The problem of reduced oxygen pressures at high altitude is also overcome by increasing the number of circulating red blood cells, so raising the oxygen carrying potential of the blood. The percentage of the total volume of the blood occupied by red blood cells in humans at different altitudes, is shown in the table:-

Altitude	Haematocrit %
sea level	47
4,500m	60

Oxygen dissociation curves for haemoglobin from subjects living at sea level and at 4,540m

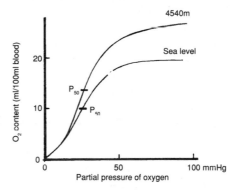

The increased oxygen content of the blood at high altitudes results from the increased number of red blood cells.

At the same time there is a decrease in the oxygen affinity of the blood at high altitude, which is reflected in the increase of P_{50}. This is a consequence of the increased intracellular concentration of 2,3-DPG.

Carriage of carbon dioxide

Carbon dioxide released by cells during respiration, is transported in the blood to the lungs. The blood which can carry approximately 2.5 times more carbon dioxide than oxygen, does this in three ways:-

- Mostly as bicarbonate ions.

- Some combined directly with haemoglobin in the form of carbamino-haemoglobin.

- Dissolved in the blood plasma.

Carbon dioxide combines with water to form carbonic acid (H_2CO_3). This reaction is considerably

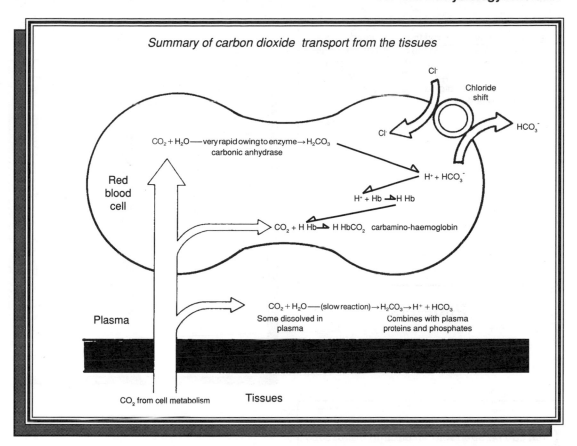

Summary of carbon dioxide transport from the tissues

accelerated in the red blood cells by carbonic anhydrase. The carbonic acid then, in turn, ionizes into hydrogen ions and bicarbonate ions. These reaction products must be removed in order that further bicarbonate can be formed. Most of the hydrogen ions are buffered by combining with deoxyhaemoglobin, while the bicarbonate ions diffuse out of the red cells into the plasma in exchange for chloride ions. The latter process is known as the chloride shift and occurs rapidly due to a chloride-bicarbonate exchange carrier system in the red blood cell membrane. These reactions are reversed in the lungs.

Carbon dioxide dissociation curves for whole blood can be determined. The position of the curve is dependent upon the oxygen content of the blood. The higher the degree of oxygenation of the blood, then the lower is its capacity for carrying carbon dioxide.

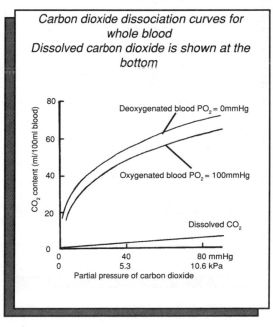

Carbon dioxide dissociation curves for whole blood
Dissolved carbon dioxide is shown at the bottom

It should be remembered that considerably more carbon dioxide is held in the blood than the small amounts that are exchanged between the tissues and lungs on each trip around the circulation. The volumes of carbon dioxide found in the blood are given in the table:-

Method of transport	Volume of CO_2 per 100ml of blood		
	Arterial blood	Additional CO_2 from tissues	venous blood
bicarbonate	42 +	2.8 =	44.8 ml
carbamino-haemoglobin	3 +	0.7 =	3.7 ml
dissolved	3 +	0.5 =	3.5 ml
TOTAL			52 ml

Platelets

Platelets are small, round or oval, non-nucleated cells.

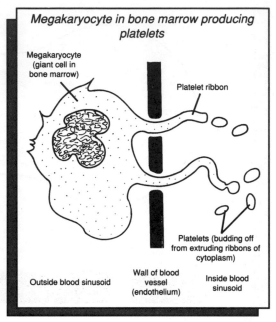

Megakaryocyte in bone marrow producing platelets

Megakaryocyte (giant cell in bone marrow)

Platelet ribbon

Platelets (budding off from extruding ribbons of cytoplasm)

Outside blood sinusoid

Wall of blood vessel (endothelium)

Inside blood sinusoid

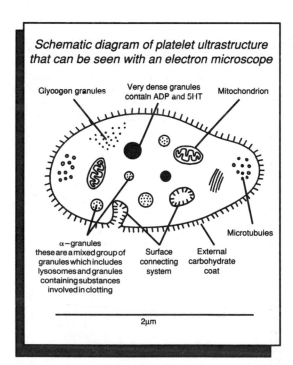

Schematic diagram of platelet ultrastructure that can be seen with an electron microscope

Glycogen granules

Very dense granules contain ADP and 5HT

Mitochondrion

α−granules these are a mixed group of granules which includes lysosomes and granules containing substances involved in clotting

Surface connecting system

External carbohydrate coat

Microtubules

2μm

Blood clotting

Platelets are important in preventing blood loss, or haemorrhage, after damage to blood vessels. The spontaneous arrest of bleeding also depends on the blood vessels themselves; various plasma proteins (clotting factors) and agents released from the damaged tissues.

There are two stages in the formation of a blood clot. The first is temporary, causes bleeding to stop within 2 to 6 minutes, and lasts about an hour. It involves both the contraction of blood vessel walls and the plugging of the leak with platelets. The permanent second stage occurs within 10 minutes of injury, and involves the production of a thrombus at the injury site (A thrombus is a blood clot occurring within the circulation). This final stable blood clot is composed of a meshwork of sticky fibrin threads that hold blood cells.

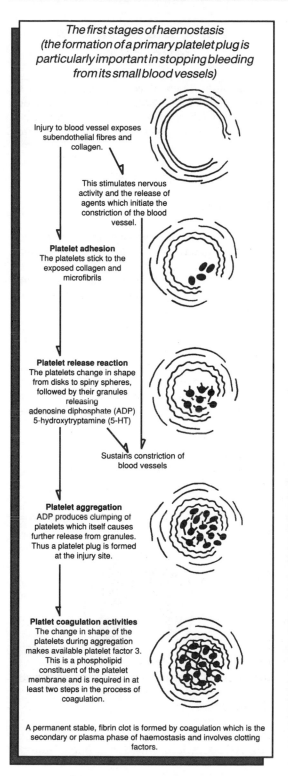

The first stages of haemostasis (the formation of a primary platelet plug is particularly important in stopping bleeding from its small blood vessels)

Injury to blood vessel exposes subendothelial fibres and collagen.

This stimulates nervous activity and the release of agents which initiate the constriction of the blood vessel.

Platelet adhesion
The platelets stick to the exposed collagen and microfibrils

Platelet release reaction
The platelets change in shape from disks to spiny spheres, followed by their granules releasing
adenosine diphosphate (ADP)
5-hydroxytryptamine (5-HT)

Sustains constriction of blood vessels

Platelet aggregation
ADP produces clumping of platelets which itself causes further release from granules. Thus a platelet plug is formed at the injury site.

Platlet coagulation activities
The change in shape of the platelets during aggregation makes available platelet factor 3. This is a phospholipid constituent of the platelet membrane and is required in at least two steps in the process of coagulation.

A permanent stable, fibrin clot is formed by coagulation which is the secondary or plasma phase of haemostasis and involves clotting factors.

Secondary or plasma phase of haemostasis

The clotting process is initiated by two different mechanisms:-

• Clotting is triggered when the blood comes into contact with an abnormal surface. For example, blood will clot when placed into a glass container. This contact activates components already present in the blood, and so this mechanism is referred to as the intrinsic pathway.

• The addition of substances which are not normally present in the blood, will also trigger clotting. Extracts from many tissues, particularly brain, will cause clotting when added to plasma. This pathway to clotting is called the extrinsic pathway.

The intrinsic and extrinsic pathways are both required for normal clotting. They converge on a final common pathway which leads to the formation of the clot.

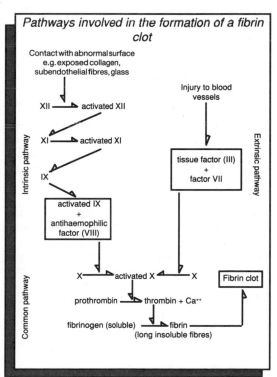

Pathways involved in the formation of a fibrin clot

The clotting pathways consist of a chain of reactions each of which produces an enzyme which will promote the next reaction in the series. This is a typical cascade process. The advantage of a cascade process, with each reaction producing a catalyst for the next, is that only very small amounts of the initial factors are required. The numerous steps in the process mean that there is considerable amplification which yields a large quantity of the final product. It assures a rapid response to damage of blood vessels.

All the clotting factors are required for proper clotting to take place. Indeed if there is a deficiency of a single protein in one of the pathways, severe bleeding disorders result. For example, haemophiliacs lack factor VIII (antihaemophilic factor), and will bleed excessively following minor injuries.

A deficiency of vitamin K, whether caused by inadequate diet, malabsorption or drugs, can also cause clotting disorders. Vitamin K is essential for the synthesis of prothrombin and other clotting factors. Another substance necessary for blood clotting is calcium (factor IV). Calcium ions are involved in several of the reactions, including the conversion of prothrombin to thrombin. The removal of calcium by chelating agents is often used as a means of preventing clotting in test tubes.

Control of clotting

The coagulation of blood must be rapid and yet not spread beyond the injury site. The balance between haemorrhage and thrombosis must be carefully controlled so that undesirable clotting does not occur. This is achieved by a number of protective mechanisms. Circulating in the blood are inhibitors that inactivate many of the activated clotting factors. One of these is antithrombin III which inactivates thrombin by forming an irreversible complex with it. The liver helps to control clotting by clearing activated clotting factors. Fibrinolysis also removes small fibrin clots by the proteolytic degradation of fibrin which is brought about mainly by the enzyme, plasmin.

White blood cells or leucocytes

The different types of white blood cell can be distinguished by their appearance in stained blood films. In the following descriptions, when colour is mentioned, stained preparations are being referred to.

Granulocytes

Neutrophils

Neutrophils are the most numerous white blood cells. Their nuclei, which are divided into between 2 and 5 lobes, stain a deep purple while their cytoplasm stains a light pink and contains granules. These granules are small and vary in size, and are probably lysosomes.

Neutrophils are phagocytes and are important in the first line of defence against bacterial invasion of the body (see chapter 4). They are highly motile, moving in an amoeboid fashion, and migrate quickly to areas of infection. During such infections the number of neutrophils in the blood increases rapidly due to mobilisation from the bone marrow. Here approximately 50% of the mature neutrophils are stored until needed. Neutrophils also help to clear the body of debris such as dead cells and thrombi.

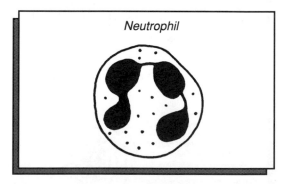
Neutrophil

Eosinophils

Eosinophils usually have nuclei with 2 lobes and their cytoplasm is crowded with bright reddish-brown granules. They are actively motile and their preferential location is near epithelial surfaces. However they are only mildly phagocytic, and do not play a direct part in the defence against bacterial invasion. They are associated with allergic reactions, since their concentration rises in such situations. They also have a role in the removal of parasites such as worms.

Basophils

The nuclei of basophils are similar to those of

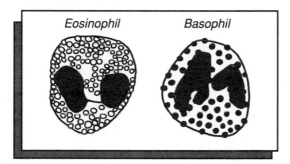

neutrophils, but are often obscured by large violet or blue-black granules. These granules contain histamine and they are released from the cell (degranulation) during allergic reactions. Basophils also contain heparin and 5-hydroxytryptamine (5HT). This, together with the fact that large numbers of basophils appear during the healing phase of inflammation, suggests that they may have a role in the prevention of coagulation of blood or lymph in obstructed tissues. Like the other granulocytes, they are extremely motile. However basophils are not phagocytic and their precise role is unknown. Their structure resembles that of the mast cells which are found in the tissues of the skin, lungs and gastrointestinal tract, but basophils and mast cells are distinct cell groups.

Monocytes

Monocytes are the largest white blood cells. Their nuclei are slightly indented into 2 or 3 lobes, and stain blue-violet. Their cytoplasm contains only a few granules and stains a hazy light blue. The monocyte moves from the bloodstream into the tissues where it matures into a macrophage. Indeed the number of macrophages in the tissues

may be 400 times the number of circulating monocytes. The tissue macrophages are given special names depending on their location, so:-

* microglial cell - brain

* Kupffer cell - liver

* alveolar macrophage - lung

* macrophage - lymph nodes, spleen

Once in the tissues macrophages may divide further in order to increase their numbers. They accumulate at the sites of lesions in inflammatory conditions and also help in immune responses. However their main function is to be actively phagocytic, and they act as scavengers ingesting bacteria, foreign bodies, damaged host cells, tumour cells and other debris.

Lymphocytes

Lymphocytes are the second most numerous type of white blood cell and are usually just larger than the red blood cell. Their nuclei are rounded or slightly indented and fill most of the cell. The lymphocyte nuclei stain deep purple while their cytoplasm stains clear light blue. Lymphocytes are the central participants in the immune system (see chapter 4).

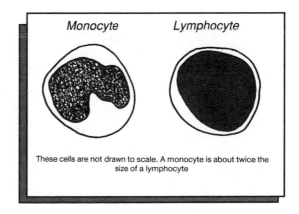

These cells are not drawn to scale. A monocyte is about twice the size of a lymphocyte

THE IMMUNE SYSTEM

Before the immune system comes into play, the body is protected from invaders by the skin and other epithelial linings acting as a physical barrier (the first line of defence). The effectiveness of this barrier is improved by various mechanisms which are shown in the diagram.

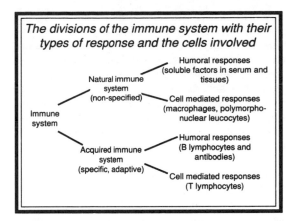

The divisions of the immune system with their types of response and the cells involved

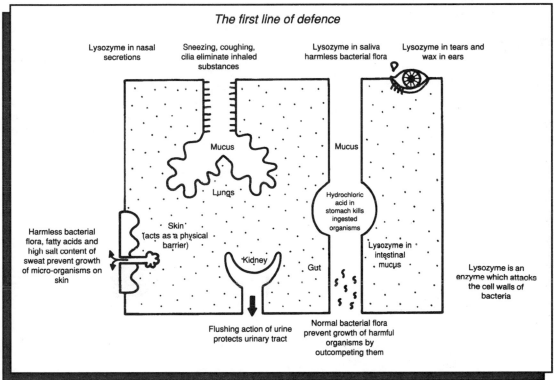

The first line of defence

Lysozyme in nasal secretions

Sneezing, coughing, cilia eliminate inhaled substances

Lysozyme in saliva harmless bacterial flora

Lysozyme in tears and wax in ears

Mucus

Mucus

Lungs

Hydrochloric acid in stomach kills ingested organisms

Harmless bacterial flora, fatty acids and high salt content of sweat prevent growth of micro-organisms on skin

Skin (acts as a physical barrier)

Kidney

Gut

Lysozyme in intestinal mucus

Lysozyme is an enzyme which attacks the cell walls of bacteria

Flushing action of urine protects urinary tract

Normal bacterial flora prevent growth of harmful organisms by outcompeting them

If the first line of defence has been breached, then the second line of defence (the immune system) is activated. The immune system protects the body against invasion by undesirable agents such as microscopic organisms, viruses, tumour cells and parasites. This protection involves the recognition and disposal of any foreign or "non-self" material.

The immune system can be divided into two parts, the natural immune system and the acquired immune system.

The natural immune system is able to recognise and respond immediately to any foreign cell or particle. This general response is always available. In contrast the response of the acquired immune system is more complex, and requires time to be fully developed. The important properties of the acquired immune system, specificity and immunological memory, are illustrated by the body's reaction to a common infectious disease. A person very rarely suffers twice from a disease such as measles. The second time that a person encounters measles, the body's defences have already been specifically activated and so prevent the disease taking hold. However this person is still susceptible to other infections such as mumps.

The responses of the immune system are divided into two types, humoral and cell mediated. Humoral responses are effected by elements free in the serum or body fluids, while cell mediated responses involve cells directly eliminating the invaders.

The natural immune system

Cell mediated response

The effector cells of the natural immune system are phagocytes, such as macrophages and polymorphonuclear leucocytes (neutrophils). An inflammatory response involving these cells occurs at the invasion site, and is characterised by heat, redness, swelling and pain. These symptoms result from local capillary dilation, slowing of the blood flow and exudation of phagocytic cells and serum bactericidal factors. If this initial inflammatory response does not prevent the spread of the foreign material, further phagocytic cells (macrophages) in the lymph nodes may succeed in eliminating it from the tissue fluids. Foreign substances entering the blood stream are dealt with by circulating phagocytes (e.g. monocytes) and by fixed macrophages in the liver and spleen.

Humoral response

• Interferons

These are proteins that are released from virus infected cells, and block virus replication in other cells. The cells receiving interferons are also protected against other types of viruses. This provides the main process of recovery from viral infections. Interferons are also produced by macrophages and T-lymphocytes.

Phagocytes taking part in an inflammatory response

Macrophages and neutrophils phagocytosing bacteria. The pus that forms is composed of masses of these phagocytes stuffed with bacteria

Dirty splinter introducing pathogenic bacteria into connective tissue

Skin epithelium

Neutrophils escaping through vessel walls

Small blood vessels which are dilated and congested with blood cells

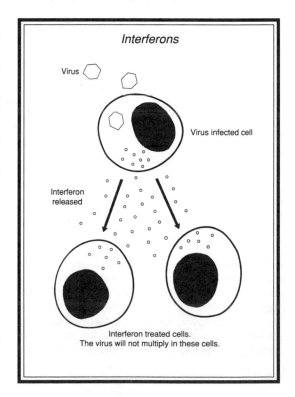

Interferons

Virus

Virus infected cell

Interferon released

Interferon treated cells.
The virus will not multiply in these cells.

• Transferrin

Bacteria are dependent on iron for multiplication, and have to compete for their iron with the iron-binding proteins such as transferrin. Hence bacteria do not thrive well in serum containing low levels of iron, but high levels of transferrin.

• Complement

This is a series of enzymes in the serum, which, when activated, produce widespread inflammatory effects and lysis of cells. Complement may be activated by certain bacteria or by antibody reactions (see later).

The acquired immune system

If the invaders overcome the natural immune system, then the acquired immune system comes into play. The humoral response of this system involves B-lymphocytes and antibody production, while the cell mediated response involves T-lymphocytes. The B-lymphocytes and T-lymphocytes are the two basic types of lymphocytes.

Following an initial encounter with an antigen, the blood is induced to produce antibodies. If later a second dose of the same antigen occurs, the concentration of the antibody is now much higher and builds up much faster than on the first occasion. This enhanced secondary response is the basis of immunological memory.

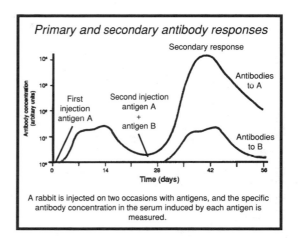

Primary and secondary antibody responses

A rabbit is injected on two occasions with antigens, and the specific antibody concentration in the serum induced by each antigen is measured.

Antibodies are proteins (immunoglobulins) which are manufactured in response to the presence of a foreign invader. Each antibody specifically recognises only that foreign substance which induced its formation. Those substances which are able to induce the synthesis of antibodies, are known as antigens.

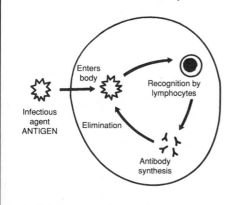

The humoral immune response

Each lymphocyte is only able to recognise and respond to one or a few closely related antigens. Lymphocytes recognise antigens when they bind to receptor molecules located on the cell surface. This binding of the antigen to the lymphocyte, activates the lymphocyte which then undergoes proliferation and differentiation.

The B-lymphocytes differentiate to become plasma cells. The plasma cells manufacture antibodies and release them into the blood and other body fluids.

Activated T-lymphocytes differentiate in several different ways. Each of the new cell types is still called a T-lymphocyte, but they have different effector functions.

T- and B-lymphocytes appear to be identical when viewed under the microscope. However there are differences in the composition of their cell membranes. B-lymphocytes, for example, have surface immunoglobulins which are receptors for antigens.

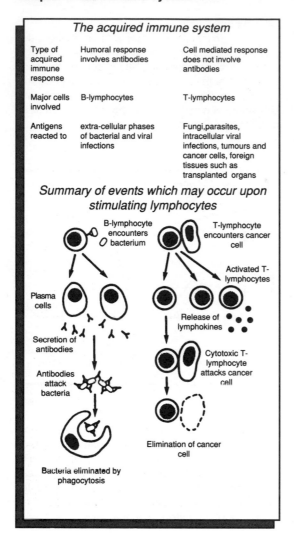

The acquired immune system

Type of acquired immune response	Humoral response involves antibodies	Cell mediated response does not involve antibodies
Major cells involved	B-lymphocytes	T-lymphocytes
Antigens reacted to	extra-cellular phases of bacterial and viral infections	Fungi, parasites, intracellular viral infections, tumours and cancer cells, foreign tissues such as transplanted organs

Summary of events which may occur upon stimulating lymphocytes

B-lymphocyte encounters bacterium

T-lymphocyte encounters cancer cell

Activated T-lymphocytes

Plasma cells

Release of lymphokines

Secretion of antibodies

Cytotoxic T-lymphocyte attacks cancer cell

Antibodies attack bacteria

Elimination of cancer cell

Bacteria eliminated by phagocytosis

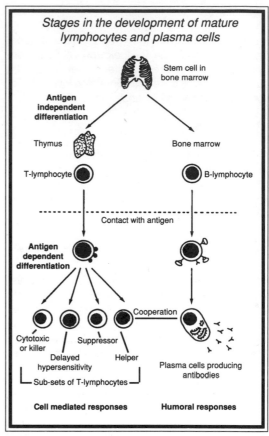

Stages in the development of mature lymphocytes and plasma cells

Stem cell in bone marrow

Antigen independent differentiation

Thymus

Bone marrow

T-lymphocyte

B-lymphocyte

Contact with antigen

Antigen dependent differentiation

Cytotoxic or killer

Delayed hypersensitivity

Suppressor

Helper

Cooperation

Plasma cells producing antibodies

Sub-sets of T-lymphocytes

Cell mediated responses

Humoral responses

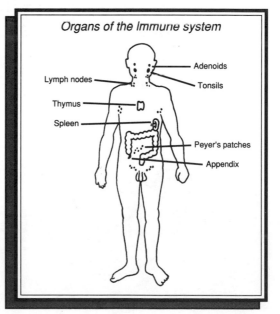

Organs of the Immune system

Lymph nodes

Adenoids

Tonsils

Thymus

Spleen

Peyer's patches

Appendix

Organs of the immune system

All lymphocytes are descended from stem cells which reside in the bone marrow (see chapter 3). Those lymphocytes destined to become T-lymphocytes, then mature during a period of residence in the thymus gland, while those destined to become B-lymphocytes mature within the bone marrow. When appropriate contact with an antigen occurs, both types of lymphocytes become activated and differentiate further. This takes place in the organs of the immune system such as the lymph nodes and spleen.

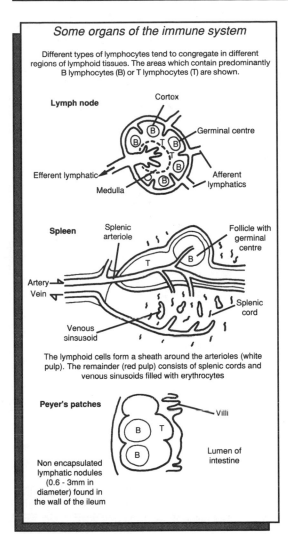

Some organs of the immune system

Different types of lymphocytes tend to congregate in different regions of lymphoid tissues. The areas which contain predominantly B lymphocytes (B) or T lymphocytes (T) are shown.

Lymph node

Cortex
Germinal centre
Efferent lymphatic
Medulla
Afferent lymphatics

Spleen

Splenic arteriole
Follicle with germinal centre
Artery
Vein
Splenic cord
Venous sinusoid

The lymphoid cells form a sheath around the arterioles (white pulp). The remainder (red pulp) consists of splenic cords and venous sinusoids filled with erythrocytes

Peyer's patches

Villi
Lumen of intestine

Non encapsulated lymphatic nodules (0.6 - 3mm in diameter) found in the wall of the ileum

The organs of the immune system are so placed that they can deal with antigens that are introduced in different parts of the body.

Organ	Source of antigens
Lymph nodes	Lymph from local tissues
Spleen	Blood
Peyer's patches	Gastrointestinal tract
Appendix	Gastrointestinal tract
Adenoids	Upper respiratory tract
Tonsils	Upper respiratory tract

The lymphatic system enhances the efficiency of the immune system in a number of ways.

• It transports antigens to the organs of the immune system.

• The continual movement of lymphocytes exposes an antigen to a large sample of lymphocytes. In this way an antigen quickly encounters any lymphocytes that can specifically react with it.

• The effector T-lymphocytes and antibodies are spread throughout the blood stream and tissues.

• It disperses the "memory" lymphocytes, in readiness for any second encounter with the antigen.

The thymus

The thymus lies in the thorax immediately beneath the upper part of the sternum. It is a pinkish-grey mass consisting of two lobes. Relative to body size, it is largest during fetal life and in the young child (at birth it weighs 10-15g). It continues to grow until puberty when it weighs 30-40g. From then on, it slowly atrophies, although it is still present in old age.

A section through a piece of the thymus showing samples of the cells found in each region

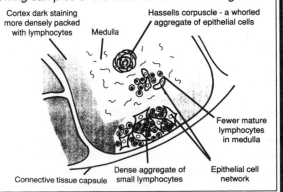

Cortex dark staining more densely packed with lymphocytes
Medulla
Hassells corpuscle - a whorled aggregate of epithelial cells
Fewer mature lymphocytes in medulla
Epithelial cell network
Dense aggregate of small lymphocytes
Connective tissue capsule

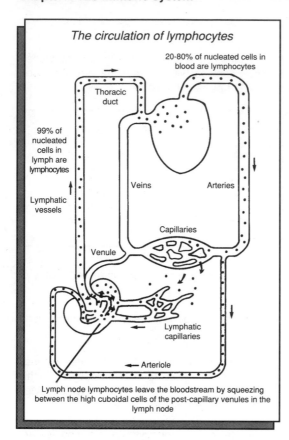

The circulation of lymphocytes

20-80% of nucleated cells in blood are lymphocytes

Thoracic duct

99% of nucleated cells in lymph are lymphocytes

Lymphatic vessels

Veins

Arteries

Capillaries

Venule

Lymphatic capillaries

Arteriole

Lymph node lymphocytes leave the bloodstream by squeezing between the high cuboidal cells of the post-capillary venules in the lymph node

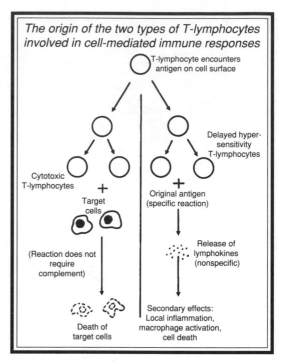

The origin of the two types of T-lymphocytes involved in cell-mediated immune responses

T-lymphocyte encounters antigen on cell surface

Delayed hyper-sensitivity T-lymphocytes

Cytotoxic T-lymphocytes

Target cells

Original antigen (specific reaction)

(Reaction does not require complement)

Release of lymphokines (nonspecific)

Death of target cells

Secondary effects: Local inflammation, macrophage activation, cell death

Cell mediated response

The differentiated effector T-lymphocytes may be divided into four distinct groups; "helper", "suppressor", "delayed hypersensitivity" and "cytotoxic" or "killer" T-lymphocytes. The cells of the latter two groups are the effector cells in cell mediated immunity. They interact directly with antigens and bring about their elimination.

The "delayed hypersensitivity" T-lymphocytes are responsible for a slowly appearing chronic inflammation in the skin. This inflammation is first visible 1-2 days following the introduction into the body of an antigen, with a peak in the intensity after 2-3 days. The development of this local chronic inflammation, is the result of the release of soluble substances called lymphokines from the "delayed hypersensitivity" T-lymphocytes. Lymphokines are a heterogeneous group of polypeptides with molecular weights from 20,000-150,000, and include inter-feron, macrophage chemotactic factor, lymphotoxin and macrophage activating factor. They have diverse properties, but in general they facilitate the elimination of the antigen by stimulating macrophages. Lymphokines are only released after a second contact with an antigen. An initial contact must have previously activated the T-lymphocytes to produce "delayed hypersensitivity" T-lymphocytes. However, once released, lymphokines are not antigen specific.

The "cytotoxic" T-lymphocytes combine directly with target cells to bring about their destruction. The target cells could include foreign cells from grafted or transplanted tissues, tumour cells and cells containing live microorganisms such as viruses. There are specific receptors on the surface of T-lymphocytes which bind to antigens on these target cells. The target cells are then lysed without the use of complement. The "cytotoxic" T-lymphocyte is not destroyed by this reaction and may continue to kill additional target cells.

The "helper" T-lymphocytes may be divided into those cells which enhance the production of "cytotoxic" T-lymphocytes, and those cells which are

Antibodies

Antibodies belong to a class of proteins called immunoglobulins (Ig). Immunoglobulin contains four polypeptide chains, two identical low molecular weight chains ("light" chains) and two identical high molecular weight chains ("heavy" chains). Both the heavy and light chains can be divided into constant and variable regions. The amino acid sequences in the constant regions of the different immunoglobulins are very similar, while the arrangements of the amino acids in the variable regions are distinct for each immunoglobulin. This variability allows for a vast number of unique antibodies each with specific affinities for different antigens. The constant regions are involved in the effector functions of antibodies. Immunoglobulins can be divided into classes based on differences in the constant regions of their heavy chains. There are five types of heavy chains μ, α, δ and ϵ, and these give rise to the five classes of immunoglobulins IgM, IgG, IgA, IgD and IgE respectively. Each of these classes has different effector functions as shown in the table. There are two types of light chain κ and λ. Both types of light chain are found in antibodies belonging to any one class, although the light chains in each individual antibody are identical.

Basic four chain structure of an immunoglobin molecule

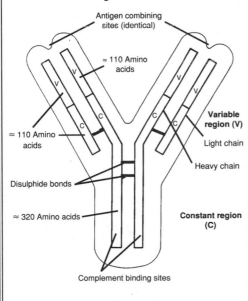

Antigen combining sites (identical)

$\approx$ 110 Amino acids

$\approx$ 110 Amino acids

Variable region (V)

Light chain

Heavy chain

Disulphide bonds

$\approx$ 320 Amino acids

Constant region (C)

Complement binding sites

essential for the formation of antibodies from B-lymphocytes. The "suppressor" T-lymphocytes also influence B-lymphocytes, but they inhibit antibody release.

Properties of human antibody classes

IgG
Heavy chains γ
Serum concentration(mg/ml) 7-16
Serum half life (days) 25

Produces major antiviral, antibacterial and antitoxin activity in serum. Stimulates ingestion by macrophages and polymorphs. Fixes complement. Crosses placenta.

IgM
Heavy chains μ
Serum concentration(mg/ml) 0.5-2
Serum half life (days) 5

Produced early in immune response. Has antibacterial and antipolysaccharide activity. Stimulates ingestion by macrophages. Good agglutinator. Fixes complement

IgA
Heavy chains α
Serum concentration(mg/ml) 1.4-4
Serum half life (days) 6

Found in body secretions including mucus, saliva, tears, milk and colostrum. Has antiviral and antibacterial activity. Protects against infections entering via eyes, naso-pharynx, urinary tract, etc.

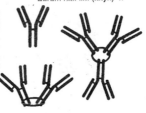

IgD
Heavy chains δ
Serum concentration(mg/ml) 0 - 0.4
Serum half life (days) 3

Present on lymphocyte surface.

IgE
Heavy chains ϵ
Serum concentration(mg/ml) 0.00002 - 0.00045
Serum half life (days) 2

Binds to mast cells in tissues. Plays role in allergies such as hayfever, and in parasitic infections.

Humoral response

Plasma cells synthesise and secrete antibodies. Any one plasma cell only manufactures antibodies of a single kind (i.e. those which are specific for the antigen that caused the initial activation of the B-lymphocyte from which that plasma cell was derived). The activated B-lymphocytes also differentiate to produce modified B-lymphocytes known as memory cells. These increase the number of lymphocytes which will respond to the antigen, and ensure a more rapid production of antibodies if the body is re-exposed to that antigen.

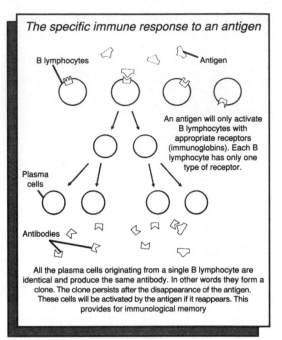

The specific immune response to an antigen

B lymphocytes — Antigen

An antigen will only activate B lymphocytes with appropriate receptors (immunoglobins). Each B lymphocyte has only one type of receptor.

Plasma cells

Antibodies

All the plasma cells originating from a single B lymphocyte are identical and produce the same antibody. In other words they form a clone. The clone persists after the disappearance of the antigen. These cells will be activated by the antigen if it reappears. This provides for immunological memory

The function of antibodies

The combination of antibodies with antigen either directly inactivates the antigen, or initiates secondary interactions which also lead to the inactivation of the antigen. The following reactions result from antigen-antibody binding:-

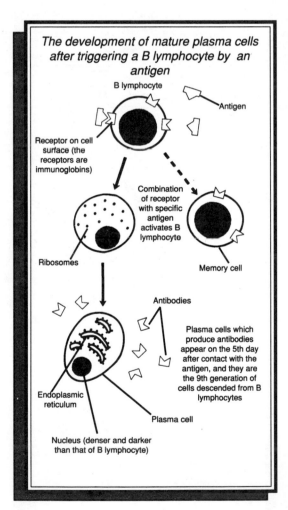

The development of mature plasma cells after triggering a B lymphocyte by an antigen

B lymphocyte

Antigen

Receptor on cell surface (the receptors are immunoglobins)

Combination of receptor with specific antigen activates B lymphocyte

Ribosomes

Memory cell

Antibodies

Plasma cells which produce antibodies appear on the 5th day after contact with the antigen, and they are the 9th generation of cells descended from B lymphocytes

Endoplasmic reticulum

Plasma cell

Nucleus (denser and darker than that of B lymphocyte)

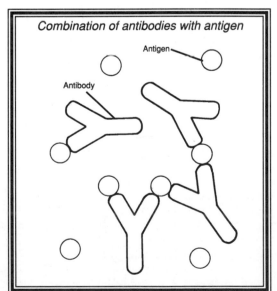

Combination of antibodies with antigen

Antigen

Antibody

• Some antigens may be directly neutralised or inactivated by the combination of the antibody with antigen. Destructive foreign enzymes such as snake venoms are inactivated in this way. The combination of antibodies with some viruses prevents them attaching to their target cells. Antibodies, in particular IgM, are able to agglutinate antigens. In this reaction the antibody crosslinks antigens to form clumps. This occurs when the antigens are invaders such as foreign blood cells or bacteria (Antibodies can act as crosslinkers because they are bifunctional reagents i.e. they have two binding sites for the antigen).

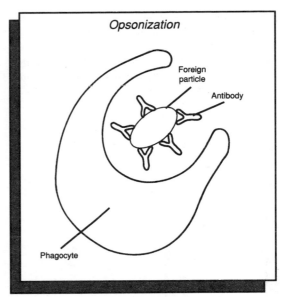

Opsonization

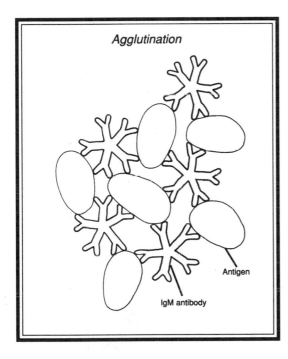

Agglutination

tions in aiding the immune response. C3, for example, promotes phagocytosis, and other complement proteins are important to the inflammatory response.

• Antibodies may attach to and coat foreign particles. This is called opsonization and targets the foreign particles for ingestion by phagocytes.

• When antibodies combine with antigens, the antibody-antigen conjugate can activate a system of soluble proteins known as complement. Triggering this system initiates a "cascade" of reactions where the final products are C8 and C9. C8 and C9 cause holes to appear in the cell membrane of the foreign invader, and this results in its destruction. Other complement components also have func-

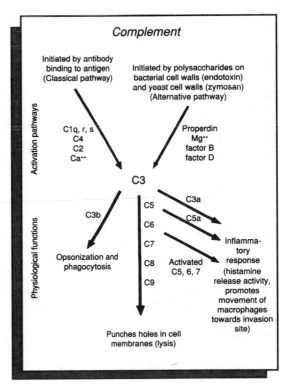

Complement

Although only B-lymphocytes can differentiate into plasma cells which synthesise antibodies, the presence of "helper" T-lymphocytes are essential for this to happen. The "suppressor" T-lymphocytes are also important as they prevent excessive antibody production.

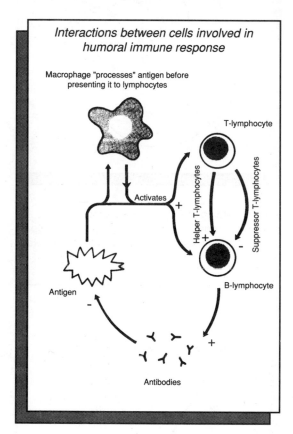

Interactions between cells involved in humoral immune response

Macrophage "processes" antigen before presenting it to lymphocytes

BLOOD GROUPS

Human blood can be divided into different types according to the presence or absence of certain mucopolysaccharides on the surfaces of the red blood cells. These mucopolysaccharides will stimulate the production of antibodies if they are introduced into an individual whose cells do not carry them. Hence the mucopolysaccharides act as antigens.

ABO system

The ABO blood group system classifies blood into types A, B, O and AB, depending on whether the red blood cells have the surface antigens A, B, neither A nor B, or both A and B. The A and B antigens are also found in saliva and other body fluids in persons with those particular red cell antigens. Individuals lacking A or B antigens normally possess antibodies belonging to the IgM class, against the missing antigen.

The ABO blood group system

Blood group	Antigen on red blood cells	Antibody in serum		Frequency of blood groups in UK and USA
A	A	anti-B		42 %
B	B	anti-A		9 %
O	-	anti-A + anti-B		46 %
AB	A + B	-		3 %

The distribution of the blood groups differs among different races. Hence Australian Aborigines are 66% blood group A, while pure Peruvian indians are 100% blood group O.

The ABO blood group system is important clinically, because blood transfusions must use blood compatible with the recipient's blood. If blood from an incompatible group is transfused, the cells clump together in large masses (agglutinate), which can have severe effects on the patient.

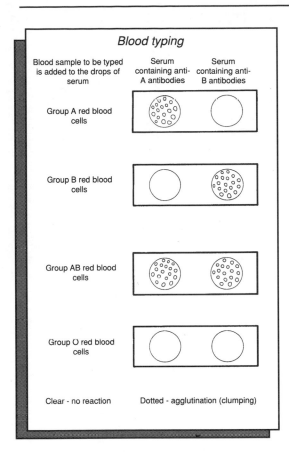

Blood typing

A large number of other human blood group systems exist. Indeed over 400 specific red blood cell antigens have been recognised to date, and it is highly unlikely that the red blood cells of any two people, have exactly the same combination of surface antigens. Most of these other red blood cell antigens do not elicit transfusion reactions like the A and B antigens.

Rhesus system

The rhesus or "Rh" blood group system is responsible for an incompatibility response between a mother and her new baby (haemolytic disease of the newborn). Eighty five percent of the population in the U.S.A. and Europe are rhesus positive (Rh⁺) and have Rh antigens on their red blood cells. Only 15% are rhesus negative (Rh⁻). Rhesus negative blood does not usually contain any anti-Rh antibodies, unlike in the ABO system where O group blood naturally contains anti-A and anti-B antibodies.

Haemolytic disease of the newborn occurs when a mother is rhesus negative, and the father and child are rhesus positive. The first time the rhesus negative mother has a rhesus positive baby, some of the baby's red blood cells may enter the mother's circulation via placental bleeds at birth. The woman then develops anti-Rh antibodies. These antibodies are of the IgG type which can cross the placenta. When the mother is pregnant a second time with a rhesus positive baby, these anti-Rh antibodies can enter the infant's circulation and destroy the infant's red blood cells causing severe haemolytic anaemia. This problem is now normally combatted by preventing the production of anti-Rh antibodies in the mother. Within 72 hours of the birth of a rhesus positive baby, the mother is given an injection of serum containing anti-Rh antibodies. This acts as a blocking agent. It interferes with the mother's reaction to her infant's rhesus positive red blood cells, and prevents her developing anti-Rh antibodies.

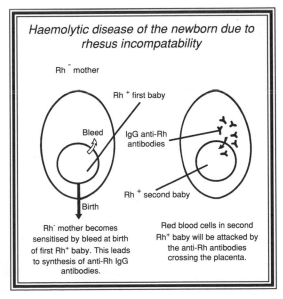

Haemolytic disease of the newborn due to rhesus incompatability

RESPIRATION

GENERAL INTRODUCTION

The lungs are part of a complex mechanism for making sure that tissues and organs are adequately supplied with oxygen, and that carbon dioxide is removed from the body. This mechanism, in which the lung function is intimately bound up with that of the heart and circulation, is controlled by the central nervous system.

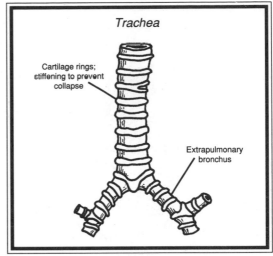

Trachea

Cartilage rings; stiffening to prevent collapse

Extrapulmonary bronchus

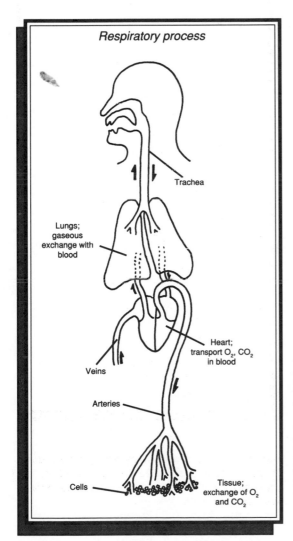

Respiratory process

Trachea

Lungs; gaseous exchange with blood

Heart; transport O_2, CO_2 in blood

Veins

Arteries

Cells

Tissue; exchange of O_2 and CO_2

LUNG STRUCTURE

Organs and tissues obtain oxygen from outside the body. This oxygen is brought into intimate contact with blood over a vast surface area. The basic structure of a lung consists of many branching tubes leading to innumerable minute air sacs, having thin, membranous walls each with a very profuse blood supply, exposed to air on both sides. The blood flow is kept moving through the lungs by the heart, and a constantly renewed supply of air is given by the repetitive "bellows-like" action of the thorax, controlled by nerve impulses from the central nervous system.

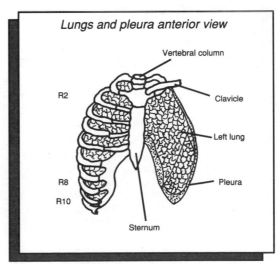

Lungs and pleura anterior view

Vertebral column

Clavicle

R2

Left lung

Pleura

R8

R10

Sternum

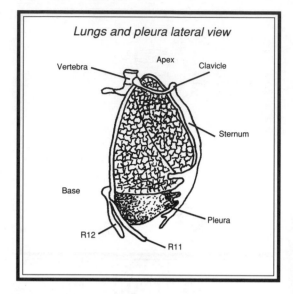

Lungs and pleura lateral view

Apex
Vertebra
Clavicle
Sternum
Base
Pleura
R12
R11

Respiratory volumes

At rest the minute volume of respiration is around 10 litres per minute. What is the volume of each breath (tidal volume)? You can check that your resting respiratory rate is about 20 per minute (provided you have not just been running for a bus) and therefore the tidal volume will work out at about 500 ml.

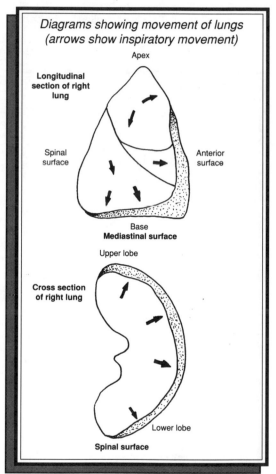

Diagrams showing movement of lungs (arrows show inspiratory movement)

Apex

Longitudinal section of right lung

Spinal surface

Anterior surface

Base
Mediastinal surface

Upper lobe

Cross section of right lung

Lower lobe

Spinal surface

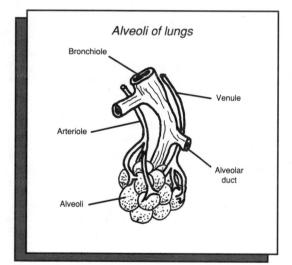

Alveoli of lungs

Bronchiole
Venule
Arteriole
Alveolar duct
Alveoli

Respiratory failure

Failure of respiration may arise from deficient blood supply or air supply, from lung defects, or from problems in the nervous control. Because of the close interrelationship of the circulation, heart and lungs, diseases of the one affect the other. When there are lung problems such as fibrosis, the right side of the heart is often dilated.

Resting respiratory rate varies with age; it is around 70 per minute at birth, 25 per minute when five years old, 20 per minute at fifteen and 15 per minute from the age of 25.

Defences of air passages

Air from outside the body is inhaled through the upper respiratory tract i.e. the warm, moist nasal cavity, larynx, trachea, bronchi and finally the respiratory epithelium of the lungs. Alveolar air is the gas which is in contact with the respiratory epithelium. Because there is an intermingling of inhaled and alveolar air with each breath, any sudden change in the composition of alveolar air is prevented.

Foreign matter, that might be harmful does not easily enter the lower respiratory tract. Even when pulmonary ventilation rises above 100 litres per minute, micro-organisms do not reach the trachea. The vestibules in the nose are provided with hairs and mucus that intercept any airborne particles. Ciliary activity induces a current out of the lungs, which has a velocity of around 2 cm/min. Various things interfere with ciliary activity, for example sulphur dioxide in the air with a concentration as low as three parts per million, or nicotine and tar in tobacco smoke.

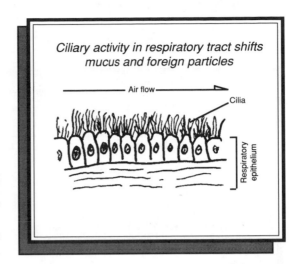

Ciliary activity in respiratory tract shifts mucus and foreign particles

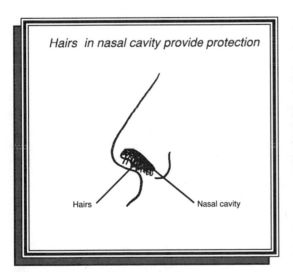

Hairs in nasal cavity provide protection

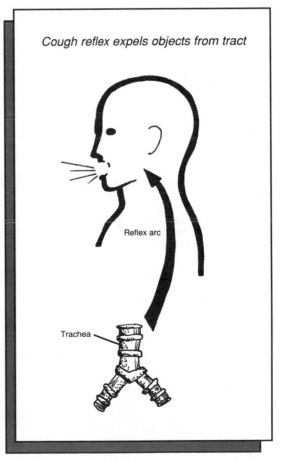

Cough reflex expels objects from tract

The cough reflex comes into play when the sensory endings in the respiratory tract are irritated by chemical or mechanical factors. The larynx and bifurcation of the trachea are the most sensitive regions, and particulate foreign matter lodged there is removed when a cough sends a rapid blast of expired air sweeping out the respiratory tree.

Volume of air in lungs

Respiratory terminology was standardised in 1950.

Lung volumes and capacities

• Vital capacity (VC) is the volume that can be expelled during maximum expiration following maximum inspiration.

• Expiratory reserve volume (ERV) is the volume expelled by a maximum expiration following a normal expiration.

• Inspiratory reserve volume (IRV) is the maximum volume that can be inhaled after a normal inspiration.

• Residual volume (RV) is the volume of air that remains in the lung after a maximum expiration.

• Functional residual capacity (FRC) is the volume of air remaining in the lung at the end of a normal expiration.

• Tidal volume (V_T) is the air shifted during one respiratory cycle.

• Total lung capacity (TLC) is the sum of IRV, V_T, ERV and RV.

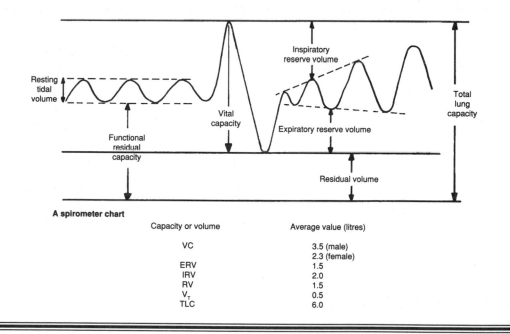

A spirometer chart

Capacity or volume	Average value (litres)
VC	3.5 (male)
	2.3 (female)
ERV	1.5
IRV	2.0
RV	1.5
V_T	0.5
TLC	6.0

Dilution of incoming air in the lung

Total volume of air entering (or leaving) the lungs in one inspiration (or expiration) is called the tidal volume and is about 500 ml at rest. About 350 ml of this atmospheric air gets to the alveoli, and mixes with air already there. The volume of respiratory gas inhaled not taking part in gaseous exchange is called the dead space. Thus, the dead space is the volume of gas not in contact with respiratory epithelium in the alveoli.

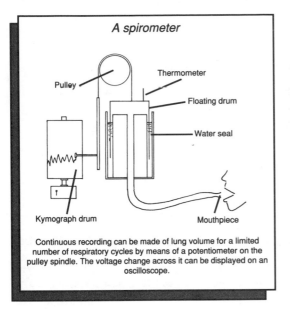

A spirometer

Continuous recording can be made of lung volume for a limited number of respiratory cycles by means of a potentiometer on the pulley spindle. The voltage change across it can be displayed on an oscilloscope.

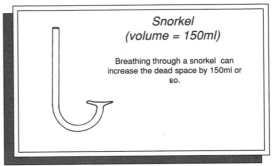

Snorkel
(volume = 150ml)

Breathing through a snorkel can increase the dead space by 150ml or so.

Physiological dead space

The physiological dead space is that space within the respiratory tract which, immediately prior to expiration, is filled with atmospheric air unmixed with alveolar air. It is air that has not taken part in gaseous exchange.

The volume of the dead space may be calculated by subtracting the volume of alveolar air expired in a single breath from the volume of all the air expired in that breath. One may calculate the volume of expired alveolar air by using the ratio of concentration of carbon dioxide in expired air to that in alveolar air.

For example, in a single breath the total volume of expired air was 486 ml and its CO_2 concentration was 4%. The CO_2 concentration in alveolar air at the same time was found to be 6%. Therefore the volume of the dead space was 162 ml, since,

$$486 \times (4/6) = 324ml \text{ and } 486 - 324 = 162ml.$$

Anatomical dead space

The anatomical dead space is the part of the respiratory tract not having alveoli; i.e. the part from the mouth to the terminal bronchioles. This volume is variable during life, because airways are not rigid. During inspiration, for example, the respiratory tubes are lengthened and dilated.

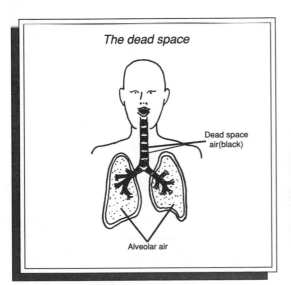

The dead space

Dead space air(black)

Alveolar air

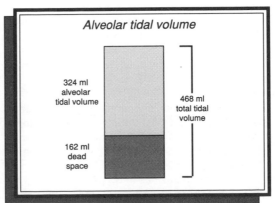

Alveolar tidal volume

324 ml alveolar tidal volume

162 ml dead space

468 ml total tidal volume

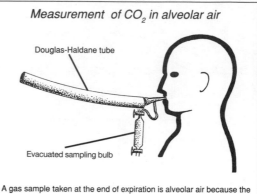

Alveolar air samples

Alveolar air is normally sampled using a Douglas-Haldane tube, which consists of a mouthpiece connected to a tube 1 metre long and 2 or 3 cm in diameter. A small side arm is attached to the tube through which the sample is taken as required. This is a sample of alveolar air because the dead space is increased, i.e. the interface between alveolar air and expired air occurs distal to the sampling tube attached to the side arm. Only a small number of breaths should be taken, otherwise there will be time for CO_2 to accumulate in the lungs and circulation and, hence, alter the composition of alveolar air.

Measurement of CO_2 in alveolar air

Douglas-Haldane tube

Evacuated sampling bulb

A gas sample taken at the end of expiration is alveolar air because the dead space is larger

Pulmonary ventilation (V$_E$)

Pulmonary ventilation, which is the same as volume of expired air per minute, is given by tidal volume x respiratory rate. Under resting conditions this is 6 to 8 litres/min. Either rapid shallow respiration or slow deep respiration will give the same V_E.

One third of tidal air remains in the dead space: the remainder reaches the alveoli, and mixes with air already there. The amount of each of these volumes varies with the depth and rate of respiration. The ratio of pulmonary ventilation to absorption of oxygen is about 3 litres per 100ml.

Alveolar ventilation (V$_A$)

The important factor in respiratory function is alveolar ventilation; it determines how much oxygen will be available for the tissues. Alveolar ventilation differs from pulmonary ventilation being a good deal less because of the dilution that occurs due to the dead space. How may alveolar ventilation be found?

Under resting conditions a subject has a dead space of 150 ml, a tidal volume of 500 ml, and a respiratory rate of 16 per minute. His alveolar ventilation would be 5,600ml/min, as compared with a pulmonary ventilation of 8,000ml/min. If now his respiration increases to 32 per minute and his tidal volume drops to 250 ml, his pulmonary venti-

lation will remain constant at 8,000 ml/min. His alveolar ventilation will now be 3,200ml/min. The dead space is a constant factor (or approximately so) in the two different situations. Hence: alveolar ventilation = respiratory rate x (tidal volume - dead space).

Taking the opposite tack, suppose the same subject now reduces his respiratory rate to 8 per minute; in order to maintain his pulmonary ventilation at 8,000ml/min, his tidal volume must be 1,000 ml. His alveolar ventilation will now be

$$8\times(1,000-150) = 6,800 \text{ ml/min.}$$

Therefore slower and deeper respiration increases alveolar ventilation; shallow respiration decreases it. In fact, dogs who lose heat from the body by

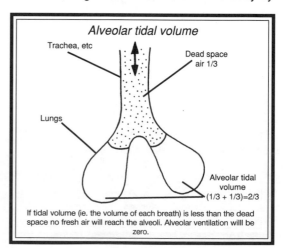

Alveolar tidal volume

Trachea, etc

Dead space air 1/3

Lungs

Alveolar tidal volume
(1/3 + 1/3)=2/3

If tidal volume (ie. the volume of each breath) is less than the dead space no fresh air will reach the alveoli. Alveolar ventilation willl be zero.

evaporation of saliva on the tongue, pant with very shallow respiration and do not increase their alveolar ventilation at all, whilst the pulmonary ventilation is enormously increased.

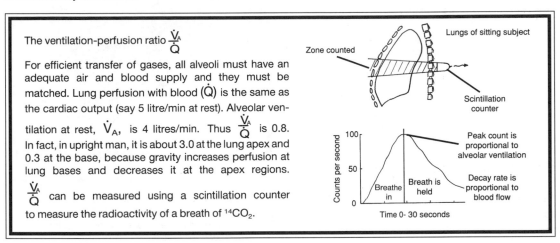

The ventilation-perfusion ratio $\frac{\dot{V}_A}{\dot{Q}}$

For efficient transfer of gases, all alveoli must have an adequate air and blood supply and they must be matched. Lung perfusion with blood ($\dot{Q}$) is the same as the cardiac output (say 5 litre/min at rest). Alveolar ventilation at rest, $\dot{V}_A$, is 4 litres/min. Thus $\frac{\dot{V}_A}{\dot{Q}}$ is 0.8. In fact, in upright man, it is about 3.0 at the lung apex and 0.3 at the base, because gravity increases perfusion at lung bases and decreases it at the apex regions.

$\frac{\dot{V}_A}{\dot{Q}}$ can be measured using a scintillation counter to measure the radioactivity of a breath of $^{14}CO_2$.

THE MECHANISM OF THE THORAX

We now have to consider the way in which air is drawn into the lungs within the thorax. Lungs can be compared with elastic balloons enclosed in a box. If air from outside enters the box, they collapse. There is a negative pressure between the balloons and the side of the box, or in the case of the lungs, between the surface of the lungs and the inside of the relatively rigid thoracic cage. It is due to the elastic recoil in the lung and can be measured. If either the lung or the thoracic wall is punctured the lungs will collapse. Therefore this is a potential space and is called the pleural cavity. In disease the pleural cavity may in fact be filled with air or fluid.

The thoracic cage is a fairly rigid framework of ribs, sternum and vertebrae, and at the bottom is the diaphragm. The capacity of the thorax is increased to cause expansion of the lungs and air is drawn in.

What would you expect to be the effect upon total thoracic volume of a sudden perforation of the chest wall? In fact there is an increase in thoracic volume. This may not have been expected but it arises from the fact that the elastic recoil of the

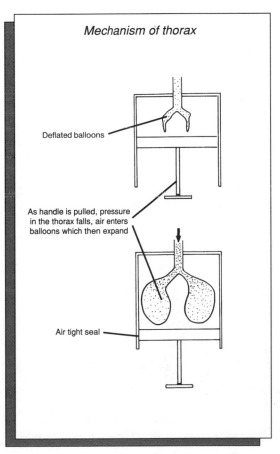

Mechanism of thorax

lungs in normal circumstances just balances the elastic expanding force in the rib cage. Air entering the pleural cavity destroys this balance and the total volume contained within the thorax is increased. The point in the respiratory cycle where the elastic recoil of the lung just balances the expanding forces in the rib cage occurs at the end of expiration. This is called the resting respiratory level. It is easy to demonstrate this on yourself by voluntarily relaxing all the respiratory muscles, when the rib cage will automatically go to this point. Notice that activity of inspiratory muscles will be required to inspire from this point or expiratory muscles will be needed to expire further.

Muscular action in rib movement

The external and internal intercostal muscles bring about inspiration and expiration by acting on the ribs to produce movements of expansion and contraction of the thoracic cage.

The external intercostals extend from the tubercles of the ribs to the costo-chondral junctions. They are thicker than the internal intercostals and the fibres slope obliquely down and forwards from the upper rib to the one below.

The internal intercostals extend from the anterior end of the intercostal space to the angles of the ribs posteriorly. Their fibres slope obliquely downwards and backwards from the upper rib to the one below.

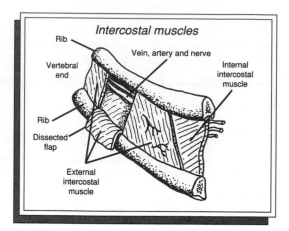

Intercostal muscles

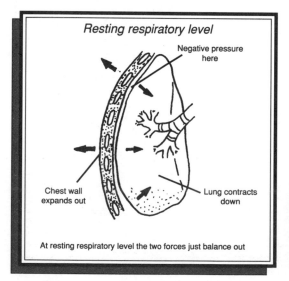

Resting respiratory level

Negative pressure here

Chest wall expands out

Lung contracts down

At resting respiratory level the two forces just balance out

The function of the intercostal muscles has been a controversial problem for a long time. During inspiration the rib cage expands because the ribs are moved around their axis of rotation. Most physiology textbooks quote this as if the ribs are on hinges or like a series of bucket handles. Their rotation increases the anteroposterior diameter of the thorax and hence its capacity. A summary of research on muscle action in breathing is as follows:-

• The mechanical action of the various intercostal muscles has not yet been definitely established. Anterior fibres raise the ribs, posterior depress the ribs, irrespective of whether they are internal or external intercostal muscles.

• The external intercostals and part of the internal intercostals contract during inspiration, both in quiet and in increased breathing.

• The same muscle groups also contract during voluntary expiratory efforts, coughing, nose blowing, etc. in conjunction with the abdominal muscles.

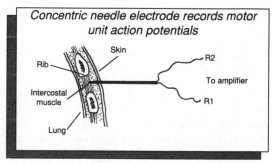

Concentric needle electrode records motor unit action potentials

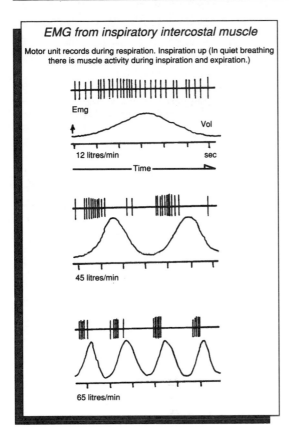

EMG from inspiratory intercostal muscle

Motor unit records during respiration. Inspiration up (In quiet breathing there is muscle activity during inspiration and expiration.)

Emg

Vol

12 litres/min   sec

———— Time ————►

45 litres/min

65 litres/min

In quiet breathing, intercostal activity is present during most of the respiratory cycle. When the whole thoracic musculature is relaxed after a normal inspiration, the expiratory movement which is due mainly to elastic recoil, happens rather too rapidly. It is therefore slowed down and controlled by muscle activity, the same muscles being active in both inspiration and expiration. Intercostals contract during normal inspiration (to produce rib elevation and to draw in air); they continue to be active, to a lesser degree, but are gradually lengthened by the elastic recoil during expiration. Only at the end of expiration do they relax completely.

This is the state of affairs up to about 50 litres/min. With moderate increases in pulmonary ventilation, up to 100 litres/min, the intercostals are active as before in inspiration, making stronger contractions; during expiration the abdominal muscles are also active.

Over 100 litres/min all the accessory muscles of respiration are brought in, e.g. sternomastoid, extensors of the vertebral column (inspiration) and all the muscles of expiration including abdominal muscles.

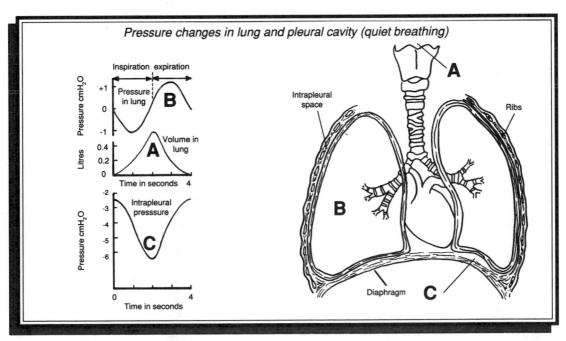

Pressure changes in lung and pleural cavity (quiet breathing)

The intrathoracic pressure

The pressure inside the pleural space can be measured simply by pushing a (sterile!) needle through the chest wall into the pleural space and watching a manometer connected to its end.

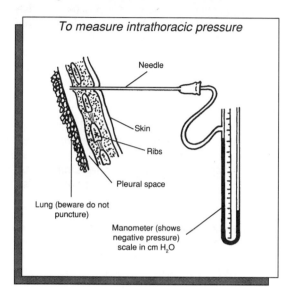

To measure intrathoracic pressure

Needle

Skin

Ribs

Pleural space

Lung (beware do not puncture)

Manometer (shows negative pressure) scale in cm H_2O

A more humane way of measuring the same pressure is by swallowing a tube with a small balloon welded to its end until the balloon lies in the oesophagus just above the cardiac sphincter of the stomach.

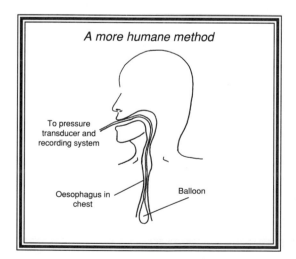

A more humane method

To pressure transducer and recording system

Oesophagus in chest

Balloon

What will intrathoracic pressure be during normal quiet breathing? In quiet respiration, intrathoracic pressure is below atmospheric pressure. This is because the lung contains elastic tissue, so arranged that it is continuously tending to collapse. Therefore, the pressure in the intrapleural (and hence the intrathoracic space) is usually negative. In quiet respiration intrathoracic pressure will vary from -3mmHg (-0.4kPa) (with respect to atmospheric pressure) during expiration, to -7mmHg (-0.9kPa) in inspiration. In certain circumstances it can become positive, i.e. above atmospheric pressure. These conditions occur when the expiratory muscles are working against the closed glottis; pressure within the lung itself rises. In coughing, for instance, intrapleural pressure may rise to +20 or +50mmHg. In defaecation and trumpet playing it may even rise to well above +300mmHg.

Movement of the lung

Lung expansion is due to elongation and dilation of the fine air passages. The alveolar ducts are most distensible. A lung does not uniformly dilate. The hilum, that is the central part, only slightly dilates; the lung surface to a depth of 3 or 4cm dilates the most. Lower lobes expand more than upper lobes. To accommodate these complicated movements, pulmonary and parietal pleura slide freely over each other.

SURFACE TENSION AND LUNG ELASTICITY

There is not enough elastic tissue within the lung substance to account for its measured elasticity. Why should this be?

Surface tension effects are responsible. In fact about half the elasticity is due to surface tension of the thin layer of tissue fluid round the inside of each alveolus. Experiments show that if you distend lung preparations alternately with air and with saline solution, you find that it takes a higher pressure if you distend the lungs with air. The interpretation of this experiment is that surface tension, which is abolished when lungs are filled with saline, is responsible. The large forces exerted by surface tension within the lung are due to the fact that the radius of a given alveolus is very small. This explains why surface tension influences the elasticity of the lung so greatly.

To measure elasticity of isolated lung

Pressure gauge

Syringe

Moist chamber   Lungs

Elasticity curves show pressures needed to expand lungs to a given volume. Lungs are more compliant when filled with saline (a) than when filled with air (b)

a   b

Tissue elasticity alone

Surface tension plus tissue elasticity

Volume of lung (% max)

Pressure (cm H_2O)

Surfactant

At functional lung volumes, the calculated effect of surface tension is 5 to 10 times too large to account for the effect described above. At larger lung volumes the measured tension comes into closer agreement. The explanation is the presence, in the lungs, of a surface active agent which reduces surface tension. These agents are called surfactants, the major component of which is dipalmitoyl phosphatidylcholine.

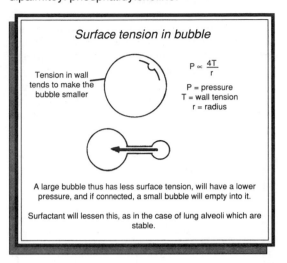

Surface tension in bubble

Tension in wall tends to make the bubble smaller

$P \propto \dfrac{4T}{r}$

P = pressure
T = wall tension
r = radius

A large bubble thus has less surface tension, will have a lower pressure, and if connected, a small bubble will empty into it.

Surfactant will lessen this, as in the case of lung alveoli which are stable.

EXCHANGE OF GASES IN THE LUNG

In man the total surface of all alveoli is 80-100sq.metres (roughly the area of a tennis court). Interchange of gas takes place between the alveolar gas and the tissue fluid. This takes place according to the fundamental gas laws.

Dalton's law of partial pressures deals with gas mixtures. It states that the pressure exerted by any gas in the mixture is the pressure that it would exert (in the same volume) if no other gases were present. Thus the total pressure of a mixture of gases is the sum of the partial pressures of the separate gases composing it. Atmospheric air when dry exerts a total pressure of 760mmHg (100kPa). 21% is oxygen, 79% nitrogen with traces of carbon dioxide and rare gases. This means that the partial pressure of oxygen (PO_2) is approximately 160mmHg (21% of 760 = 160).

Water vapour

Air in contact with water continuously receives water molecules from the interface. This vapour also follows the law of partial pressures, exerting its own vapour pressure independently of the other gases and proportional to the amount of it present. This amount varies with temperature, air at higher temperatures containing more water vapour. At 37°C (temperature in the lungs) fully saturated water vapour pressure is 47mmHg (6.25kPa). On the other hand at room temperature, 18°C, the partial pressure of water vapour is 15.5mmHg when fully saturated. Of course, at any temperature, air may not be fully saturated with water in which case the partial pressure of the water vapour will be correspondingly lower.

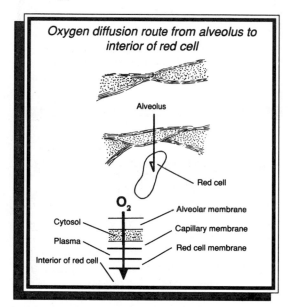

Oxygen diffusion route from alveolus to interior of red cell

Mechanism of gas exchange

Gas pressures are fairly constant in alveolar air; only slight changes occur between expiration and inspiration. Alveolar air is in equilibrium with systemic arterial blood. Composition of alveolar air is determined by several factors:-

• Composition and pressure of expired air.

• Volume of tidal air, respiratory frequency, volume of the dead space.

• Volume remaining in the lung at the end of expiration.

• Velocity of oxygen absorption by the blood.

• Rate of loss of carbon dioxide from the blood.

Physical laws are sufficient to explain gas exchange processes in the mammal. Barcroft showed that the PO_2 in alveolar air is always slightly higher than in the blood. In healthy persons the absorption of oxygen is explained by its capacity to diffuse though the lungs, although it is limited by pulmonary ventilation and particularly by the circulation though the lungs.

Carbon dioxide diffuses twenty times as fast as does oxygen. As far as carbon dioxide is concerned, a gradient of 0.03mmHg in the partial pressure of carbon dioxide (PCO_2) is enough to ensure elimination by simple diffusion of all the carbon dioxide produced by a person at rest.

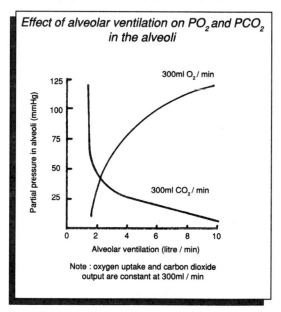

Effect of alveolar ventilation on PO_2 and PCO_2 in the alveoli

Note : oxygen uptake and carbon dioxide output are constant at 300ml / min

Pulmonary diffusing capacity

This is the number of ml/min of gas transferred across the pulmonary epithelium per unit pressure difference. To measure this value we need to know (in the case of oxygen):-

• The number of ml of oxygen transferred from alveoli to blood per minute (In other words the

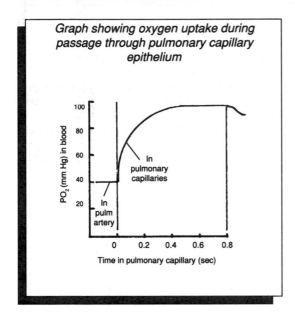

Graph showing oxygen uptake during passage through pulmonary capillary epithelium

Factors affecting diffusing capacity

Apart from partial pressure difference of the gas between air and blood, diffusing capacity is proportional to :-

1. Surface area of pulmonary vascular bed

2. Inverse of alveolar membrane thickness

3. Ease of gas diffusion through membrane

4. Solubility of gas in membrane itself

The normal value in a young adult is between 20-30 ml O_2 / min / mmHg (it reaches a maximum of 75 in exercise).

oxygen consumption per minute).

• The mean alveolar PO_2. This is difficult to measure since the distribution of oxygen throughout the alveoli is uneven.

• The mean pulmonary capillary PO_2 (This is also difficult to measure and can only be arrived at by means of an equation called Bohr's integration).

Pulmonary diffusing capacity can also be meas-

ured by the use of carbon monoxide in low concentrations; carbon monoxide has 210 times the affinity for haemoglobin than does oxygen.

We have now dealt with the factors involved in transferring oxygen from the atmosphere into the blood in contact with the alveolar membranes. The way in which blood transports oxygen and carbon dioxide is described in chapter 3.

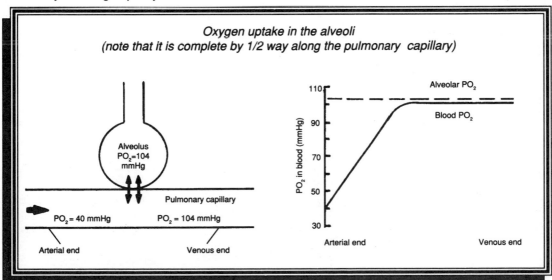

Oxygen uptake in the alveoli
(note that it is complete by 1/2 way along the pulmonary capillary)

CONTROL OF RESPIRATION

Oxygen requirement and carbon dioxide production vary, and must be controlled. The initiation of respiration is the rhythmic discharge of motor neurones in the cervical and thoracic spinal cord supplying the respiratory muscles.

The control of respiration is brought about by two neural mechanisms:-

• A voluntary system, which originates in the cerebral cortex.

• An automatic system in the pons and medulla.

The voluntary system takes care of respiratory excursions required during speech etc., whereas the involuntary system controls the overall amount of gas exchange in relation to the metabolic requirements of the body. In the case of the voluntary mechanism, nerve impulses travel to the spinal motor neurones involved in respiration via the corticospinal tract. The automatic system involves the reticular spinal pathways.

What are the mechanisms that control respiration? The first observation is that if you cut all the nerves to the respiratory muscles breathing stops. This must indicate that the respiratory muscles of the ribs and also the diaphragm do not have spontaneous rhythmicity. They must be stimulated to contract by the nervous system. Although it is possible to control breathing voluntarily for short periods of time as in talking, coughing etc., breathing is essentially an automatic rhythmical process. It continues to work even in sleep and moderately deep anaesthesia.

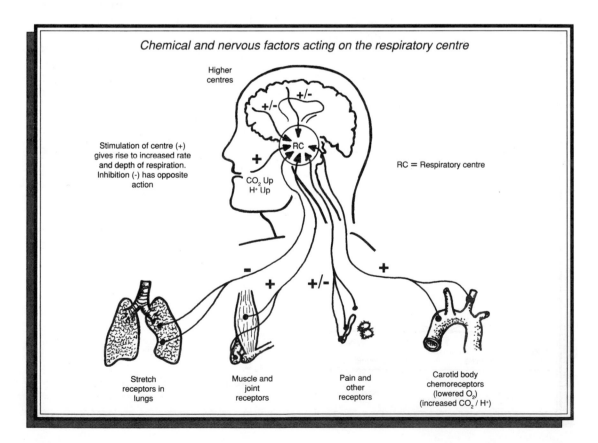

Chemical and nervous factors acting on the respiratory centre

Higher centres

Stimulation of centre (+) gives rise to increased rate and depth of respiration. Inhibition (-) has opposite action

CO_2 Up
H^+ Up

RC = Respiratory centre

Stretch receptors in lungs

Muscle and joint receptors

Pain and other receptors

Carotid body chemoreceptors (lowered O_2) (increased CO_2 / H^+)

The respiratory centre

If the brains of experimental animals are cut across at different levels, the effect on respiration can be studied. When the cut is made anterior to the pons there is no effect on respiration. If the cut is slightly posterior to this (between the medulla and the pons) respiratory rhythm is changed. If the cut is just posterior to the medulla all respiration stops. This has then been taken to mean that there is a group of nerve cells, which we may term the respiratory centre, that maintains the rhythm of respiration. Moreover the site of this so-called centre is in the medulla. It also appears as if it is modified by other regions such as the pons. Microscopic examination, stimulation experiments, and recording with microelectrodes lead to the conclusion that there are nerve cells within the medulla, some of which are resposible for inspiration, and others for expiration. It is normally reckoned that these form a diffuse cellular network.

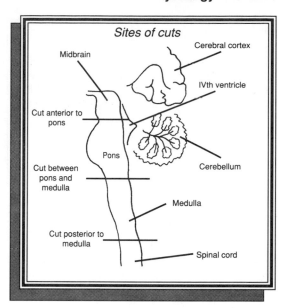

Sites of cuts

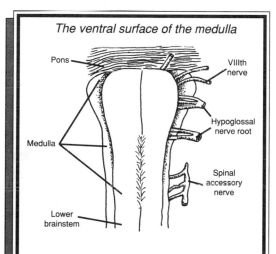

The ventral surface of the medulla

Central chemosensitivity of respiration is represented by a nervous region in the ventral surface layer of the medulla. Its natural elements are a heterogenous group of nerve cells having a tonic irregular firing pattern. Their degree of activity is influenced by the pH of the interstitial fluid around them.

The cells are localised in three bilateral areas close to the ventral surface of the medulla, extending from the root of the hypoglossal nerve to the border of the lower brainstem.

Respiratory rhythm

There has been considerable argument for many years as to how rhythmical bursts of action potentials for the production of breathing are generated. There is no agreement even now. It appears that the respiratory neurones briefly activate the inspiratory muscles. That is followed by a period in which the neurones are relatively quiet. Passive expiration occurs unless expiratory neurones are active. It would therefore appear that the medulla has some inherent rhythmicity contained within it. Moreover this inherent rhythmicity can be modified by chemical and neural inputs.

Recent work, however, tends to emphasize the importance of spinal reflexes in maintaining a large part of respiratory rhythm.

Changes in breathing

The respiratory system can operate over a wide range. Both depth and rate are subject to control. Depth of respiration is basically controlled by the density of impulses from respiratory neurones to respiratory muscles. The higher the frequency of impulses and the larger the number of motor units activated, the stronger are respiratory contractions;

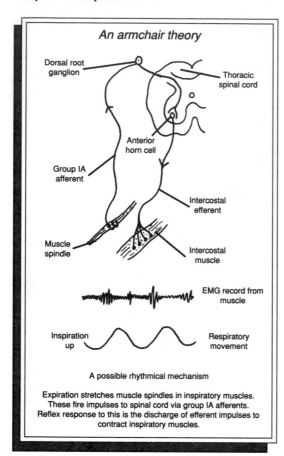

An armchair theory

Dorsal root ganglion

Thoracic spinal cord

Anterior horn cell

Group IA afferent

Intercostal efferent

Muscle spindle

Intercostal muscle

EMG record from muscle

Inspiration up

Respiratory movement

A possible rhythmical mechanism

Expiration stretches muscle spindles in inspiratory muscles.
These fire impulses to spinal cord via group IA afferents.
Reflex response to this is the discharge of efferent impulses to
contract inspiratory muscles.

Peripheral chemoreceptors

The chemoreceptors are stimulated mainly by a fall in the partial pressure of oxygen in arterial blood. The receptors initiate nerve impulses and these are transmitted via afferent nerve pathways to the medulla, stimulating it to give a reflex increase in pulmonary ventilation. These receptors are active when there is a low partial pressure of oxygen in the inspired air, such as at high altitude. They are also stimulated by an increase in acidity.

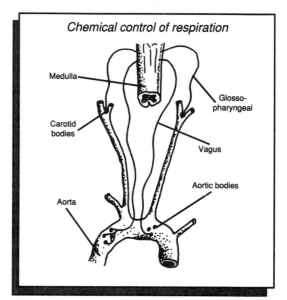

Chemical control of respiration

Medulla

Glosso-pharyngeal

Carotid bodies

Vagus

Aortic bodies

Aorta

this is the same kind of control mechanism as occurs in ordinary skeletal muscle. The control of respiratory rate is a function of the rate at which these bursts of action potentials are discharged by the medullary neurones. By and large this is controlled by chemical inputs. In exercise the body requires more oxygen and gives off more carbon dioxide and this therefore requires an increase in respiratory gas exchange. As one might expect, a deficiency of oxygen in the blood and an excess of carbon dioxide (or hydrogen ions) stimulate respiration.

Two mechanisms are responsible for this.

• Chemoreceptors are situated in the arch of the aorta and in the carotid body region, for blood.

• Chemoreceptors sense the chemical content of the cerebrospinal fluid within the brain.

Chemoreceptors in the medulla

It used to be thought that the respiratory centre itself in the medulla was sensitive to partial pressure of carbon dioxide. In fact, evidence now shows that there are separate chemoreceptors which are anatomically close to the respiratory cells in the medulla and these relay nerve inpulses to the centre. The most powerful stimulant here is a rise in the partial pressure of carbon dioxide in the cerebrospinal fluid. This reflexly causes an increased respiratory rate and depth. In the opposite sense, when the carbon dioxide partial pressure in the blood drops below normal levels, ventilation may actually cease until the level rises to normal once again. A period of voluntary hyperventilation results in the washing out of CO_2, the consequent lowering of partial pressure of carbon dioxide in circulating blood and hence to inhibition of respiration.

The effect of carbon dioxide is a good bit more powerful in regulating respiration than is oxygen deficiency (which acts on the peripheral chemoreceptors). This is the reason for breathing not beginning again after hyperventilation until the carbon dioxide level approaches normal, even if there is a considerable lack of oxygen at the same time.

There is an effect of oxygen lack on the central chemoreceptors that we are talking about, and that is a depressive one. The general effect of oxygen lack on medullary cells of whatever function is to depress them, and the respiratory cells are no exception.

Other afferent pathways to respiratory centres

The main input to the cells which regulate breathing according to oxygen demand and carbon dioxide output comes from chemoreceptors. However, respiration is modified by other influences. The first is stretch receptors in the lungs and chest wall. At the inspiratory phase of respiration the lungs fill with air, they stretch, and receptors sited within them fire off impulses, terminating inspiration. This is known as the Hering-Breuer reflex.

There are stretch receptors in skeletal muscle and in joints which become stimulated during exercise. They send impulses to the respiratory neurones in the medulla. This can be shown experimentally by passively moving the arms or legs and it is found that respiratory rate and depth are thereby increased. Other receptors in all parts of the body, in addition, send afferent impulses to the respiratory control system. Factors such as painful stimulation of the skin, changes in skin temperature, presence of food in the mouth, etc. will have marked effects on respiration.

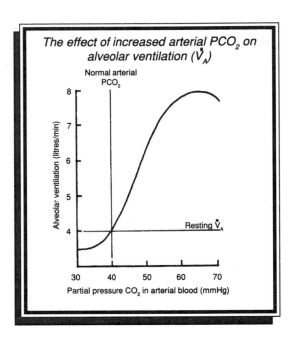

The effect of increased arterial PCO_2 on alveolar ventilation ($\dot{V}_A$)

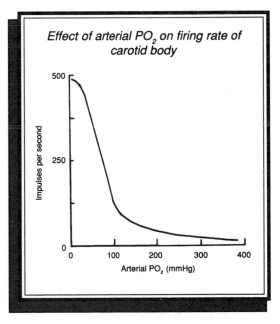

Effect of arterial PO_2 on firing rate of carotid body

THE HEART

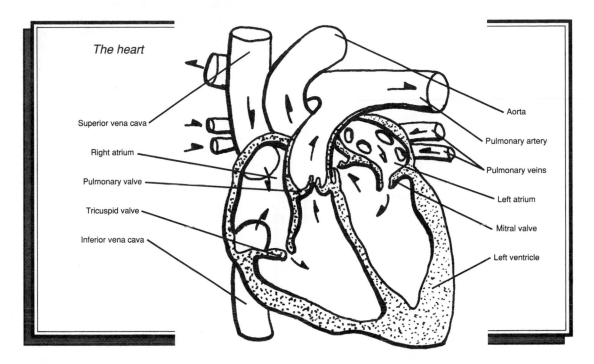

The heart

- Superior vena cava
- Right atrium
- Pulmonary valve
- Tricuspid valve
- Inferior vena cava
- Aorta
- Pulmonary artery
- Pulmonary veins
- Left atrium
- Mitral valve
- Left ventricle

The heart is an amazing structure. Throughout life it pumps a minimum of 5 litres/min into the aorta under high pressure, and in strenuous exercise this can be increased by 6 times. Heart muscle has no resting period except for the pause between each beat.

Structure

The mammalian heart is divided into two separate pumps. This arises because the pressures required to perfuse the systemic circulation are much higher than required for the lungs. The former needs to be up to 120mmHg; the latter not much more than 25mmHg. Each side is in the form of two chambers, an atrium and a ventricle. The flow of blood through the heart can be seen in the figure. From the head and the rest of the body, blood flows into the large veins; the superior and inferior venae cavae. From there the blood flows into the right atrium. Then it passes to the right ventricle, is pumped to the pulmonary arteries and while the blood is still deoxygenated it goes to the lungs. In the lungs blood becomes oxygenated and flows back to the heart via the pulmonary veins and then enters the left atrium. It flows from there to the left ventricle, and that pumps the blood into the aorta, hence circulating it through the body.

S = peak systolic pressure (mmHg)
D = diastolic pressure (mmHg)
CO = cardiac output (litres/min)
$\overline{SAP}$ = mean arterial pressure (mmHg)
SV = stroke volume (ml)
CI = cardiac index

$\overline{SAP} = S - ((S-D)/3)$
CI = CO/Body surface area
SV = CO/heartrate

$\overline{SAP}$ = 80 - 110 mmHg
CI = 2.8 - 4.2 litres/min/m²
SV = 70 - 200 ml

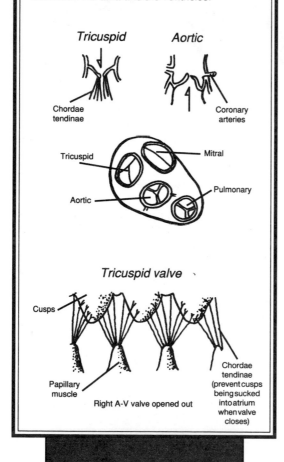

Heart valves

The valves in the heart are shown in the figure and it is important to realise that all these valves are completely passive structures. They open and close in accordance with the pressure differential across them. If the pressure is higher on the inlet side blood will flow, if pressure is higher on the outlet side the valves will fold back, shut and close off the orifice. The pulmonary and aortic valves separate the ventricles from the pulmonary artery going to the lung and the aorta going to the body, while the atrioventricular valves (A-V valves) are betweeen the atria and the ventricles.

Tricuspid   *Aortic*

Chordae tendinae

Coronary arteries

Tricuspid — Mitral

Aortic — Pulmonary

Tricuspid valve

Cusps

Papillary muscle

Right A-V valve opened out

Chordae tendinae (prevent cusps being sucked into atrium when valve closes)

Blood supply

The heart muscle might be thought to obtain its blood supply from the blood within the chambers of the heart. However, the muscle is far too thick for diffusion to supply the muscle with oxygen, therefore the muscle has a system of smaller coronary arteries to supply it with blood. The important feature of the blood supply to the heart muscle is that these are end arteries. In other words, a given zone of heart muscle is supplied by a single artery and there is no overlap. This is why blockage, by disease processes, of the blood supply to the heart is such a dangerous phenomenon.

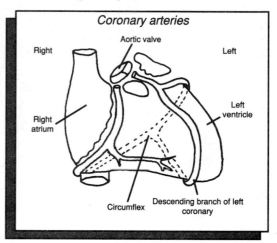

Coronary arteries

Aortic valve

Right   Left

Right atrium

Left ventricle

Circumflex   Descending branch of left coronary

Heart muscle

The properties of heart muscle are between smooth muscle and skeletal muscle. The fibres do have cross striations like skeletal muscle but are branched forming a network, so that the whole of the heart is a syncytium. When a part of the muscle is excited, impulses travel from that region throughout the ventricles or the atrium. The heart as a whole is made up of two of these fibre networks; one in the atrial walls, and the other in the two ventricles.

Nerve supply to the heart

The heart is controlled by the autonomic nervous system, both sympathetic and parasympathetic branches. Control is via the rate modulation in the parasympathetic (vagus) nerve supply to the heart. In most mammals including man there is vagal tone. This means that impulses are constantly travelling down the vagus from the cardiac centre to the heart and tending to slow it down.

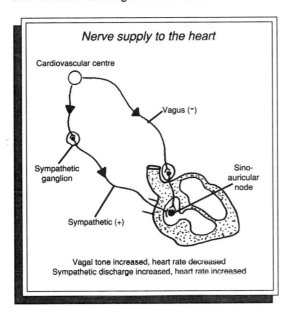

Nerve supply to the heart

Cardiovascular centre

Vagus (¬)

Sympathetic ganglion

Sino-auricular node

Sympathetic (+)

Vagal tone increased, heart rate decreased
Sympathetic discharge increased, heart rate increased

Contraction time

Cardiac muscle is not like skeletal muscle because it relaxes very slowly. The duration of contraction of ventricular muscle is about 300msec while for skeletal muscle the contraction is over within 30 to 40msec. The function of the long contraction is to enable blood to be pumped out during one contraction.

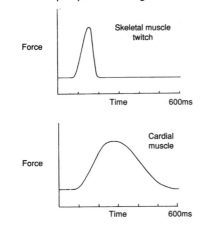

Force — Skeletal muscle twitch
Time — 600ms

Force — Cardial muscle
Time — 600ms

Refractory period

The refractory period in heart muscle is as long as the contraction and this compares with about a tenth of the contraction time in skeletal muscle. The result of this is that heart muscle cannot contract until it has fully relaxed from the previous beat. It will therefore not give a tetanic contraction. This is important because if heart muscle went into a prolonged tetanus the heart would not pump.

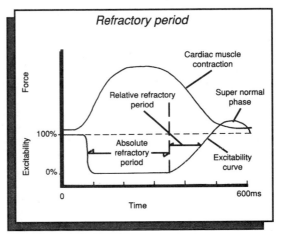

Refractory period

Force

Cardiac muscle contraction

Relative refractory period

Super normal phase

Excitability

100%

Absolute refractory period

Excitability curve

0%

0 — Time — 600ms

Conducting tissue

Heart muscle cells can all transmit impulses. However, some cells in the heart are specially modified in order to conduct impulses, and they do so at a faster rate than ordinary heart muscle fibres. There are two kinds:-

• the sinoatrial node (SA node) in the right atrium

• the atrioventricular node (AV node) between the right atrium and right ventricle

The SA node is where the impulse appears first; it initiates the heart beat (the pace-maker). It is controlled by the vagus nerve which slows the heart, and the sympathetic nerves which increase the heart rate. The terminal ramifications of these nerve fibres end in the SA node.

The AV node is also a pacemaker but does not normally have this function. If the SA node is removed experimentally or is damaged by a disease process, the AV node then takes over but the heart rate is slower than normal. It is also supplied by nerve fibres and, in the absence of the SA node, heart rate can still be altered by activity of the cardiac centre. The AV node gives rise to modified muscle called Purkinje fibres (conducting tissue) which pass down the inter-ventricular septum, and are then distributed to ventricular muscle. The main branch is called the bundle of His, and divides into right and left bundle branches. They are modified muscle cells, not nerve fibres and conduct the impulses about eight times as fast as normal cardiac muscle cells.

THE CARDIAC CYCLE

The main events of the cardiac cycle are changes in blood pressure within the cavities of the heart, changes of volume, functions of the heart valves, heart sounds and electrical changes taking place in heart muscle.

When heart muscle is undergoing a contraction this is the period of systole; during relaxation it is diastole. Atrial systole preceeds ventricular systole.

The propagation of the cardiac impulse

The SA node starts the cardiac impulse. It then travels through the atrial wall in muscle cells. There are no specialised conducting pathways in the atria. The impulse consists of a travelling depolarisation-repolarisation sequence which passes rapidly from one cell to another generating, almost at once, muscle contraction. This impulse does not travel directly to ventricular muscle because of the connective tissue ring with no muscle in it, lying between ventricles and atria. The only way the impulse can travel to the ventricular muscle is via the AV node. There is a delay of about 100msec in the AV node. The delay enables the atria to empty the blood within them into the ventricles before these contract. The AV node then becomes depolarised, the cardiac impulse passes quickly along the bundle of His, finally being transmitted to the muscle cells. The architecture of the conducting system enables the action potential to reach all the ventricular fibres at about the same time, so that the muscle contracts simultaneously.

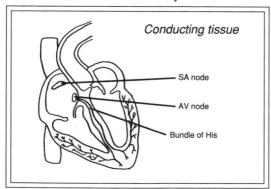

Conducting tissue

SA node

AV node

Bundle of His

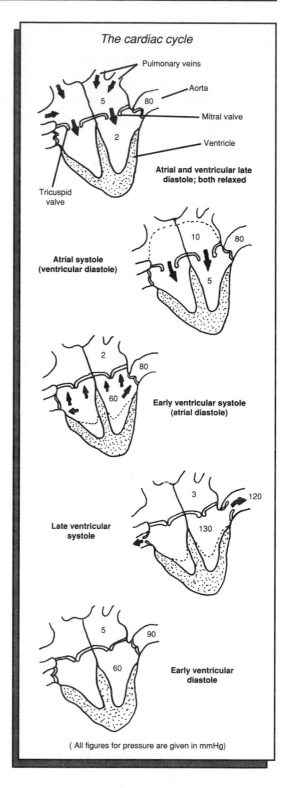

The cardiac cycle

Pulmonary veins

Aorta

Mitral valve

Ventricle

Tricuspid valve

Atrial and ventricular late diastole; both relaxed

Atrial systole (ventricular diastole)

Early ventricular systole (atrial diastole)

Late ventricular systole

Early ventricular diastole

(All figures for pressure are given in mmHg)

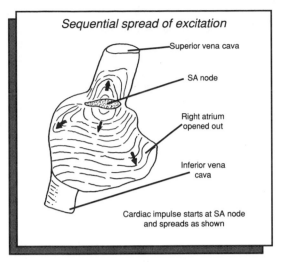

Sequential spread of excitation

Superior vena cava

SA node

Right atrium opened out

Inferior vena cava

Cardiac impulse starts at SA node
and spreads as shown

The electrocardiogram (ECG)

The fact that heart muscle generates electric currents has been known for many years and action potentials can be picked up at a distance from the heart. It is possible to find potentials of about 1mV (peak to peak) when electrodes are placed on the body extremities.

The various waves of the electrocardiogram have been labelled P, Q, R, S and T. The P wave results from depolarisation of atrial muscle. It indicates movement of the action potential across the atria. It occurs slightly before atrial contraction.

The QRS complex results from the wave of depolarisation in both the ventricles; it slightly precedes the contraction of the ventricles.

The T wave is due to the repolarisation of the ventricular muscle. The atria also repolarise, of course, but this cannot be seen because it occurs in the middle of the QRS complex. The interest of the electrocardiogram is that it enables the function of the heart to be mirrored in a recording. Abnormalities of cardiac function are shown by characteristic changes in the electrocardiogram.

Rhythmicity

If the nerve supply to the heart is blocked, the heart still continues to beat, but faster than normal because it is now following the rate of discharge of the SA node. If the SA node is removed the heart rate is slower because it is now controlled solely by the AV node. If the AV node also is removed, the heart beat is slower still, the so-called idio-ventricular rhythm. It is possible to remove from the body small pieces of heart muscle and keep them alive in an organ bath where they will display spontaneous rhythmicity without the benefit of conducting systems at all. This proves that the cardiac muscle cells themselves are inherently rhythmical.

The ECG

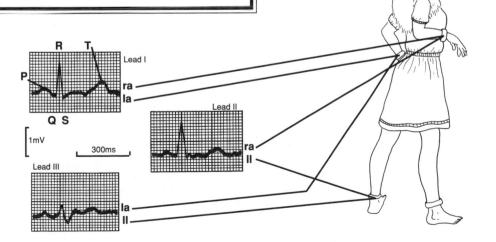

Lead I

P R T

Q S

ra
la

1mV   300ms

Lead II

ra
II

Lead III

la
II

ECG DIAGNOSIS

For diagnostic purposes, leads may be attached to the periphery of the body in a large number of places. For instance, lead I is between the right arm and left arm, lead II the right arm and left leg, and lead III left arm and left leg. If there is a disparity between the sizes of the QRS complex in these leads, it indicates a shift in the electrical axis of the heart. Left axis deviation occurs when there is a large upright QRS complex in lead I and a large inverted QRS complex in lead III. Right axis deviation is when in lead I the QRS complex is inverted and in lead III it is large and upright. The QRS complex (and the T wave) are generated as a result of opposing depolarisations in the right and left ventricles. The voltages due to each ventricle are similar but the time courses are not.

At the start of the QRS complex, right and left ventricular depolarisations sum to give rise to the R wave (since activity travels in the same direction in the interventricular septum). The downstroke of the R wave is due to cancellation of the resultant depolarisations as they progress in opposite directions, back around the outer ventricular wall. Since the left ventricle is larger than the right, its muscle fibres are longer and therefore depolarisation lasts longer. This gives rise to the T wave, i.e. it is the unopposed electrical activity in the left ventricle.

This electrical balance can be disturbed either because of hypertrophy (increase in size) of muscle fibres in one or other ventricle, or by an actual anatomical displacement of the axis of the heart. These abnormalities are termed axis deviations.

The heart rate

The ECG is an accurate method for measurement of heart rate; even during severe exercise it is possible to fix small radio transmitters to subjects and to obtain records of the response of the heart to muscle work. However, the major importance of the ECG lies in clinical diagnosis of disorders of cardiac rhythm.

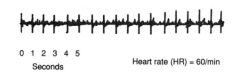

0 1 2 3 4 5
Seconds

Heart rate (HR) = 60/min

CARDIAC RHYTHMS

Abnormal rhythms often originate at an abnormal site for initiation of impulses. Characteristics of cardiac arrhythmias depend on several factors:-

• Any part of the myocardium and/or conducting system can initiate excitation.

• The heart is functionally four separate masses of muscle joined together by electrically conducting tissue.

• Excitation spreads through the heart from an excited region.

• Spread via muscle or Purkinje fibres may follow an abnormal course.

Abnormal rhythms may arise in an ectopic focus of excitation. Paroxysmal tachycardia (rapid heart rate) can originate either in atria or ventricles and the rate may rise to over 150/min. The recording of

Axis deviation

Left A.D.

I   II   III

Right A.D.

I   II   III

Records from two patients showing (top record) left axis deviation and (bottom) right axis deviation (tracings from leads I,II and III shown).
Left axis characterised by tall R wave in lead I and large S wave in lead III. This indicates a shift of the heart's electrical axis to the left, either anatomically or because of ventricular hypertrophy.
Right axis deviation is indicated by large S wave in lead I and big R wave in lead III. Chronic hypertension or aortic incompetence give s rise to left axis deviation; mitral stenosis or emphysema leads to right axis deviation.

the ECG shows regular and rapidly repeated complexes. Ventricular fibrillation consists of randomly travelling patterns of excitation in the ventricles; hence no co-ordinated muscle contraction occurs and the heart beat ceases. Survival is not possible for more than one or two minutes. This is the cause of death in some forms of drowning, electrocution and coronary heart disease.

Abnormal heart rhythms

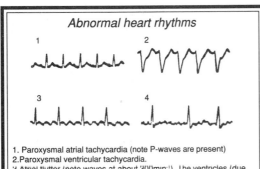

1. Paroxysmal atrial tachycardia (note P-waves are present)
2. Paroxysmal ventricular tachycardia.
3. Atrial flutter (note waves at about 300min⁻¹). The ventricles (due to their refractory period) respond to one in three atrial waves and the HR is around 100 min⁻¹.
4. Atrial fibrillation. Ventricular rhythm is irregularly irregular due to random excitation from the disorganised atrial muscle.

Coronary occlusion

The major cause of death in middle age is coronary heart disease. Pathological processes result from an infarct (death of muscle fibres) which follows occlusion of a coronary artery or one of its branches. The occlusion is usually the result of progressive narrowing by atheroma. Acute blocking of the coronary arteries produces the characteristic changes in the ECG that can be seen in the figure.

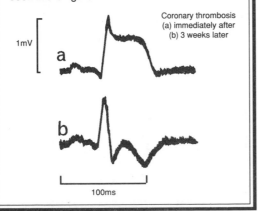

1mV

a

b

Coronary thrombosis
(a) immediately after
(b) 3 weeks later

100ms

Pressures during cardiac cycle

The blood flow in the heart is determined only by the pressure differential across the orifices of the heart, and the opening and closing of heart valves. Heart valves are passive structures opening and closing according to the pressure across them. The series of diagrams shows the pressure changes in the left side of the heart during one cardiac cycle. The pressure in the aorta does not ever fall below 80mmHg during the cardiac cycle. The reason is that the elasticity of the aortic walls plus that of the large arteries, maintains the pressure by elastic recoil and smooths out the difference between systolic and diastolic pressures.

Ventricular filling

About 70% of the filling of the ventricle is a passive function and takes place before the atrium begins contracting. Because the pressure in the aorta is about 80mmHg, and in the ventricle is about zero, the aortic valve is closed. There is no backflow into the ventricle.

Atrial systole

When the atrium contracts, presssure within it increases and the remaining 30% of the blood is forced into the ventricle because the pressure in the atrium is above that in the ventricle, the A-V valve is open and blood is flowing through it. However, the pressure increase due to atrial contraction is small because atrial muscle is thin and does not develop much force. Then the atrial muscle relaxes; atrial systole is over, and atrial diastole continues for the rest of the cycle.

Atrial diastole

During diastole of the atrium, pulmonary venous pressure is slightly greater than in the atrium, so the atrium is filling throughout diastole. As a result, atrial pressure increases slightly.

Ventricular systole

When the atrium relaxes ventricular muscle begins contraction. Pressure within the ventricle rapidly rises above atrial pressure, therefore the A-V valve suddenly shuts and no backflow takes place into the atrium.

Both A-V and aortic valves are now closed, therefore the ventricle is a closed chamber containing a given volume of blood. Pressure in this phase rises steeply as force is developed by the ventricular muscle, although there is no actual shortening of muscle fibres. This is the isometric phase of ventricular contraction. As soon as intraventricular pressure is above the 80mmHg in the aorta, the aortic valve is suddenly opened, and blood is then ejected from the ventricle into the aorta. There is a continuation of blood flow throughout the remainder of ventricular systole and the maximum pressure reached is around 120mmHg.

Then the ventricle relaxes (diastole), the pressure falling rapidly. As it does so, it reaches a level lower than that in the aorta and at this instant the aortic valve closes. Then there is an isometric phase of relaxation until the intraventricular pressure is even lower than the pressure within the atrium and at this point the A-V valve opens. There is blood flow into the ventricle and filling of the ventricle begins again. The next cycle begins with a further atrial systole.

The systole of the ventricle has a duration of about 300msec and at normal heart rates diastole is about 500msec. If heart rate increases, ventricular systole does not decrease in duration, but the duration of diastole becomes progressively curtailed as heart rate increases. This naturally poses problems because there is less time for ventricular filling when heart rate is high and the efficiency of the heart as a pump gets less. There is a point of diminishing returns and if rate rises much above about 150/min it does not increase cardiac output very much.

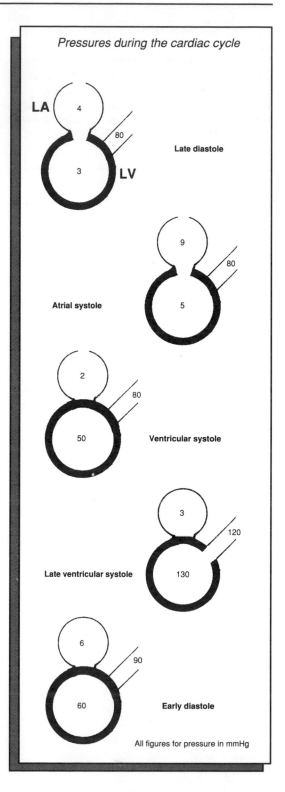

Pressures during the cardiac cycle

LA / LV

4
80
3
Late diastole

9
80
5
Atrial systole

2
80
50
Ventricular systole

3
120
130
Late ventricular systole

6
90
60
Early diastole

All figures for pressure in mmHg

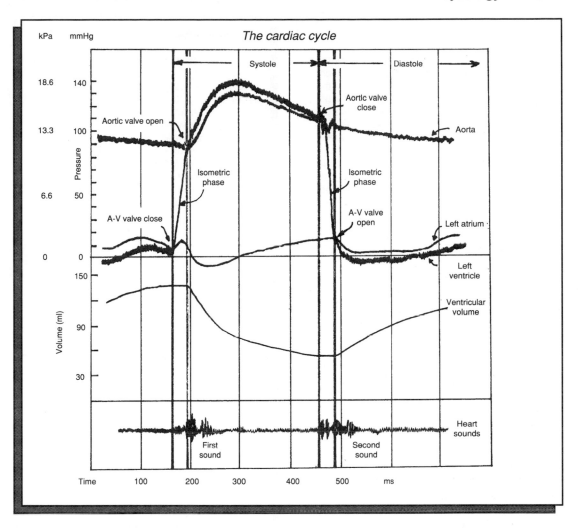

The cardiac cycle

Cardiac output

Cardiac output is the volume of blood discharged by either the right or the left ventricle per minute. This volume varies with the demand of the body for blood. The resting output of the heart is usually 5 litres/min. In exercise this figure can be increased to 25 litres/min in most people, but trained athletes may go up to 40 litres/min.

Cardiac output may be increased in two ways. Firstly, the stroke volume can be increased. Secondly, the heart rate can be increased or, of course, both changes can occur. Other things being equal, cardiac output equals stroke volume times heart rate.

$$CO = SV \times HR$$

Pressure changes in the pulmonary side of the heart

The changes taking place in the cardiac cycle are the same on the right and left sides of the heart with the difference that the pressure required to drive blood through the lungs is relatively low. Lungs do not offer the resistance to flow that is offered by the various organs in the systemic circulation and only about 25mmHg is reached in the right ventricle during its contraction. However, the same volume of blood is pumped out at each beat.

Heart sounds

The heart sounds are generated mainly by the noise of the valves closing. They may be heard using a stethoscope placed on the skin over the heart. The usual description is given as ... *lubb-dup*. ... The first heart sound arises from the closure of the two A-V valves; the second arises in the closing of the aortic and pulmonary valves. These represent the normal heart sounds, but a number of different adventitious sounds, or heart murmurs, may also be detected. These abnormal sounds are usually associated with blood turbulence generated as blood passes through leaking valves or through a valve that has a narrowed orifice resulting from disease.

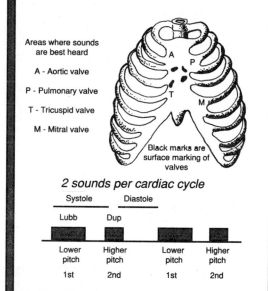

Areas where sounds are best heard

A - Aortic valve

P - Pulmonary valve

T - Tricuspid valve

M - Mitral valve

Black marks are surface marking of valves

2 sounds per cardiac cycle

Systole		Diastole	
Lubb	Dup		
Lower pitch	Higher pitch	Lower pitch	Higher pitch
1st	2nd	1st	2nd

Heart rate control

The normal resting heart rate is 72/min; it can be raised to 200/min. The factors involved in this regulation are mostly the nervous system changes in the vagus and cardiac sympathetic nerve supply. There are other mechanisms operative as well. The outflow via the nerve pathways to the heart (vagus, sympathetic) is co-ordinated by a network of neurones in the medulla. Incoming nerve impulses from structures such as pressure receptors, muscle, joints and skin, determine the outflow from this region as a feedback control system. This diffuse network of nerve cells is the cardiovascular centre. Part of it is concerned with control of heart rate; part of it is concerned with the degree of constriction of blood vessels throughout the body.

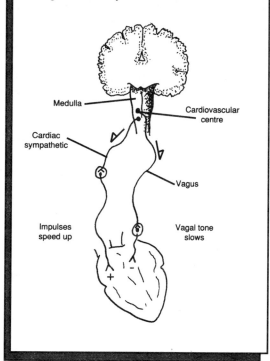

Medulla

Cardiovascular centre

Cardiac sympathetic

Vagus

Impulses speed up

Vagal tone slows

Heart rate control via the SA node

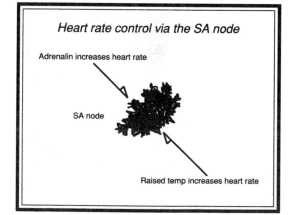

Adrenalin increases heart rate

SA node

Raised temp increases heart rate

Heart rate control via the SA node

Various hormones act directly on the SA node. For instance, adrenalin stimulates it and thus speeds up heart rate. The SA node is also responsive to body temperature. As body temperature rises, as in fever, heart rate is increased.

Stroke volume control

As we have seen, cardiac output is directly dependent upon stroke volume. In the average human it can vary between 70ml/stroke to about 250ml/stroke. The single most inportant factor that determines stroke volume is venous return. The heart puts out exactly the amount of blood that enters it from the veins. Stroke volume is determined partly by intrinsic properties of cardiac muscle, partly by nervous and chemical influences both on cardiac muscle and its conducting tissue.

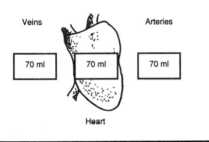

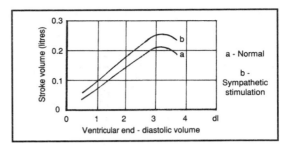

Nervous control of stroke volume

Not only do the nerve fibres going to the heart have ramifications of their endings on the SA and AV nodes but they also innervate ventricular muscle. Sympathetic fibres (SNS) elicit much stronger contraction of muscle, resulting in the emptying of the ventricles being more complete with each beat.

Normally, the heart is putting out the same volume of blood as that entering it. At rest a small increase in the volume of blood entering the ventricles gives rise to increased stretch of cardiac muscles and stronger contractions putting out the extra blood. When the venous return is increased by a large amount, the sympathetic supply to the heart also increases the strength of contraction of the ventricles and raises the heart rate.

Starling's law

All muscle fibres display a characteristic relationship between length and force. As far as skeletal muscle is concerned, this is a relationship which is at a peak at normal length of muscle in the body. At lengths both below and above normal body lengths, the force exerted by muscle falls away. This results from the mechanics of the contractile mechanism (see chapter 8). Heart muscle is normally working on the rising slope of this relationship. As the interdigitating filaments of actin and myosin are pulled out, the force produced by a contraction increases and over a short range, this relationship is substantially linear. This linear relationship is called Starling's law of the heart. This means that, within limits, when cardiac muscle fibres are stretched they contract with more force and an extra volume of blood is ejected. This is an intrinsic property of cardiac muscle and means that the ventricles eject all but a small amount of the blood entering them. Hence, a transplanted heart without nerve supply will function more or less normally, provided no excessive demands are placed on it. This intrinsic property of cardiac muscle is the most important factor in balancing the input and output of the two ventricles.

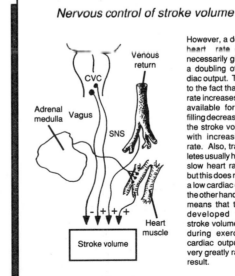

Nervous control of stroke volume

However, a doubling of heart rate does not necessarily give rise to a doubling of the cardiac output. This is due to the fact that as heart rate increases, the time available for diastolic filling decreases, hence the stroke volume falls with increasing heart rate. Also, trained athletes usually have a very slow heart rate at rest but this does not denote a low cardiac output; on the other hand, it usually means that they have developed a large stroke volume and that during exercise their cardiac output can be very greatly raised as a result.

CIRCULATION

CIRCULATION OF THE BLOOD

The varying requirements for blood by organs and tissues, are met by a high pressure, arterial reservoir of blood fitted with taps which control the run off of blood to tissues. Modification of the resistance to flow through one of these taps, i.e. in the blood supply to an organ, results in a change in flow through it. The heart provides the driving force and maintains the high pressure reservoir. However the pressure in the ventricle varies between zero and 120mmHg and would not provide a constant perfusion pressure for the organs. These pressure variations are ironed out by the large arteries and the aorta, which are highly elastic. When blood is ejected by the ventricle into the aorta, the pressure is raised; when ejection ceases, the aortic valve shuts and the elastic recoil of the, by now, stretched

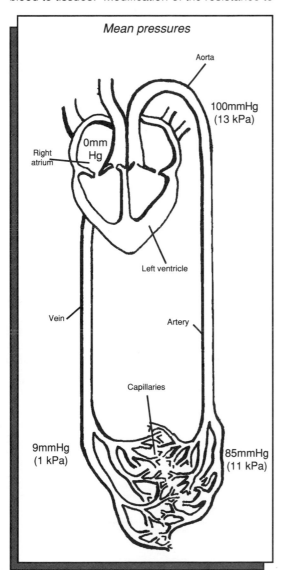

Mean pressures

Aorta

100mmHg
(13 kPa)

0mm Hg

Right atrium

Left ventricle

Vein

Artery

Capillaries

9mmHg
(1 kPa)

85mmHg
(11 kPa)

Pressure gradients

A decrease in pressure occurs between the aorta and great veins, because there is a large resistance to flow in the circulatory system. This is called the peripheral resistance. It is a frictional resistance, largely between the blood and the walls of the blood vessels. Part of the resistance is actually within the blood itself. There is a decrease in arterial pressure between the aorta and the great veins. However, it is not a smooth decrease. In the aorta and large arteries the mean blood pressure falls from, say, 100mmHg to 90mmHg, whereas a much larger fall, say from 90mmHg to 25mmHg takes place across the length of the arterioles. The greater pressure drop across the arterioles arises because there is a greater resistance to flow. This is because the radius of these vessels is small and, comparatively speaking, the velocity of flow there is high. Therefore the frictional component is large. It might be thought that the largest resistance to flow would be in the capillaries, but the total cross-sectional area of the capillaries is very large so that the velocity of flow is very small. The total cross-sectional area of all the arterioles is similar to that of the aorta and therefore the velocity in them is comparatively high.

Main pressure drop is across arterioles

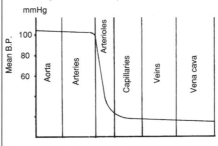

arterial walls, results in a roughly even blood flow through the systemic arteries. Blood is pumped into the aorta with a mean pressure of 100mmHg which decreases continuously as the blood flows through the body, and ultimately, when the blood comes back to the right atrium the pressure is near zero.

The pulse

The pressure in the arterial tree varies throughout the cardiac cycle. This variation, when felt at a peripheral site such as the wrist, is called the pulse. The pressure variations are more noticeable at

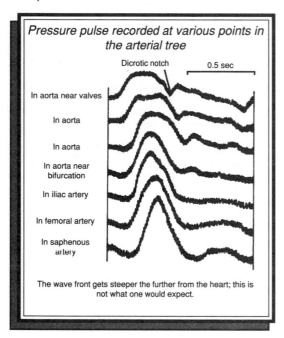

Pressure pulse recorded at various points in the arterial tree

The wave front gets steeper the further from the heart; this is not what one would expect.

some sites than others. Blood comes smoothly out of a cut vein but a cut artery gives rise to blood in spurts.

Regulation of blood flow

The systemic circulation is made up of many circulations in parallel. This leads to a sophisticated control mechanism to maintain the blood flow through a particular organ, if one of the others in parallel with it changes its demands. This control is one of the chief functions of the vasomotor centre

Velocity of flow

Since it is a closed system, blood flowing through any part of the system of arteries and veins at each minute must be the same as the cardiac output. If the cardiac output is 5 litres/min, then 5 litres flow in the aorta, in the sum total of all the arteries, in all the capillaries, in all the veins and the flow back to the heart is 5 litres/min. This, of course, is the volume of blood flow per minute. The rate of blood flow varies greatly and depends on the cross-sectional area of the vessels concerned. The aorta is a very large artery and its cross-sectional area is about 5cm². It branches into many arteries, each smaller than the aorta but having a greater total cross-sectional area. For the arterioles, the total area is slightly larger still. The area of cross-section of all the capillaries is about 1000 times that for the aorta. Therefore blood flows along a capillary at 1000th of the rate that it does in the aorta. In the aorta, velocity is about 60cm/s. In the capillaries it is 0.06cm/s. Most capillaries are of the order of 1mm in length which means that a given sample of blood will only stay there for one or two seconds whilst it is exchanging oxygen, carbon dioxide, etc.

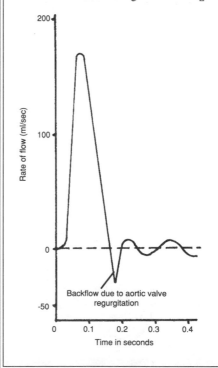

Flow in ascending aorta of dog

in the medulla. Skeletal muscle has a very variable requirement for blood. During strenuous exercise, muscles may get 90% of the blood ejected by the heart. Under these circumstances, blood flow to the kidney might be only about 2% of cardiac output. Normally the kidneys will get 20% of total blood flow. This blood distribution is controlled by the amount that arterioles constrict in different parts of the body.

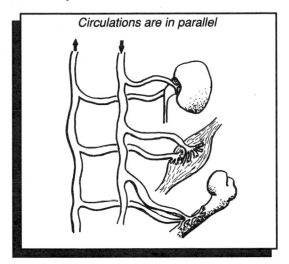

Circulations are in parallel

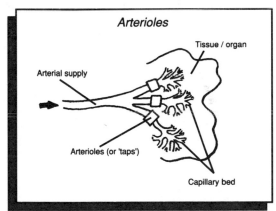

Arterioles

Tissue / organ

Arterial supply

Arterioles (or 'taps')

Capillary bed

Arteriolar tone and blood distribution

The arterioles (or resistance vessels) are the seat of the peripheral resistance. For any vessel, resistance (R) is inversely proportional to the fourth power of the radius (r) of the vessel in question.

$$R \propto \frac{1}{r^4}$$

Hence small increases in arteriolar radius give rise to large decreases in resistance. For example, an arteriole has its resistance to flow halved by a change of radius from 0.1 to 0.12mm.

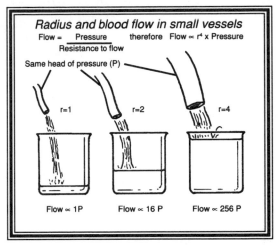

Radius and blood flow in small vessels

$$Flow = \frac{Pressure}{Resistance\ to\ flow} \qquad therefore \quad Flow \propto r^4 \times Pressure$$

Same head of pressure (P)

r=1   r=2   r=4

Flow ∝ 1P   Flow ∝ 16 P   Flow ∝ 256 P

Local dilation of arterioles

The diameter of arterioles is controlled by a local mechanism in the tissues. This is a self-regulating mechanism and it works in order to increase the blood flow when a tissue is active. An example would be contracting skeletal muscle. Any increase in activity of a tissue gives rise to increased production of metabolites, e.g. carbon dioxide or lactic acid. These have an effect directly upon the smooth muscle in the walls of the arterioles and they dilate. The resistance to flow in the active region is reduced, and the blood flow is increased.

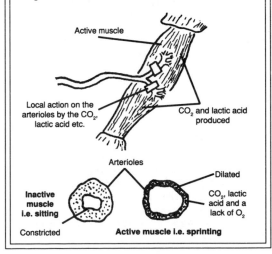

Active muscle

Local action on the arterioles by the CO_2, lactic acid etc.

CO_2 and lactic acid produced

Arterioles

Dilated

Inactive muscle i.e. sitting

CO_2, lactic acid and a lack of O_2

Constricted

Active muscle i.e. sprinting

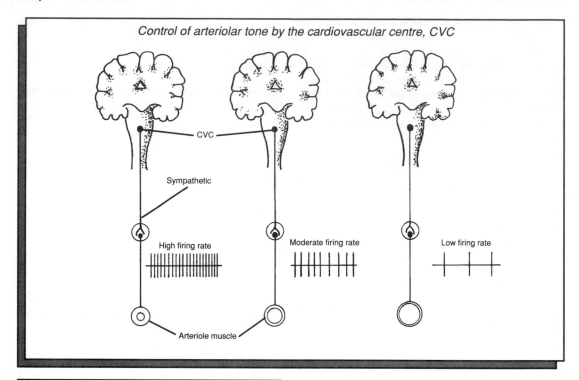

Control of arteriolar tone by the cardiovascular centre, CVC

CVC

Sympathetic

High firing rate

Moderate firing rate

Low firing rate

Arteriole muscle

Sympathetic control of arterioles

The vasomotor centre is part of the cardiovascular centre in the medulla which determines the degree of contraction of smooth muscles in the arteriolar walls. Under normal circumstances it sends a tonic (continuous low level) discharge of impulses in sympathetic nerves to the smooth muscle, keeping the vessels partly contracted. There is not a parasympathetic supply. The amount of contraction is increased by increasing the frequency of discharge of sympathetic impulses to the smooth muscle in the vessels. Conversely, when the requirement is for more blood to flow into an organ, i.e. the resistance to flow needs to decrease, then the vessels are dilated by lowering the frequency of impulses. Thus altering sympathetic vasomotor tone, changes blood flow to an organ.

There are other parts of the brain which also have effects on the tone of peripheral vasculature. The hypothalamus has effects on peripheral resistance and it mediates reactions to fear, anger or pain. Also, in body temperature regulation, the hypothalamus controls the amount of blood that flows in vessels in the skin. Obviously a profuse skin blood flow will enable heat to be transferred in the circulating blood from deeper regions to the external environment.

Hormonal effects on arterioles

The hormone, adrenaline, is concerned with the blood vasculature. Activation of the sympathetic pathway from the brain to the adrenal medulla releases adrenaline. Adrenaline acts like the sympathetic system, i.e. it constricts arteriolar smooth muscle. However, there is an exception to this: the blood supply within the muscle. Here adrenaline dilates the small vessels and promotes increased blood flow.

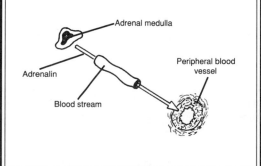

Adrenal medulla

Adrenalin

Peripheral blood vessel

Blood stream

BLOOD PRESSURE

Arterial pressure

The maintenance of a relatively constant, high, arterial blood pressure, is essential to the functioning of the circulatory system. If the pressure becomes too low there will be a lack of pressure to force blood through the small vessels. If the pressure rises too high, there is the danger of rupturing small vessels, for example in the brain. The brain especially requires a constant and copious blood supply. If it fails, consciousness will be lost. If it is cut off for more than one or two minutes, brain cells die.

The relationship between arterial pressure (BP), cardiac output and peripheral resistance (TPR) is like Ohm's law. It is $BP = CO \times TPR$. If either cardiac output or peripheral resistance rises, then arterial pressure will rise, and vice versa. Also if one of the three factors is changed, inevitably there will be repercussions in the other two.

Arterial pressure

Small vessel under low pressure

No blood flow

Small vessel under high pressure

Haemorrhage

These problems can be seen directly in the retina

During exercise cardiac output may increase about five times. If there were no change in peripheral reistance then blood pressure would rise to 500mmHg, a level which would give rise to minor explosions throughout the vascular tree. The peripheral resistance must therefore be reduced to about one fifth of its normal value, if blood pressure is to be kept at its normal level. It should be understood that in exercise, cardiac output does increase by up to five times, but only as a result of the peripheral resistance decreasing. The decrease takes place in the active muscles and, because of the decrease, cardiac output increases in order to maintain the blood pressure.

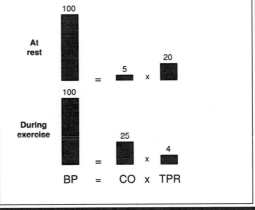

At rest	100	=	5 x 20
During exercise	100	=	25 x 4

BP = CO x TPR

Calculation of total peripheral resistance (TPR)

$$BP = CO \times TPR$$

$$\text{Therefore } TPR = BP / CO$$

For example: BP = 100mmHg
        CO = 5 litres/min

then TPR = 100/5 = 20 mmHg/litre/min

Control of arterial pressure

Blood pressure regulation is a complex procedure because of the interacting factors, blood pressure, cardiac output and peripheral resistance. It follows that pressure may be controlled by negative feedback systems operating upon cardiac output and peripheral resistance separately or both at once.

Baroceptor reflexes

The arterial pressure is controlled by negative feedback servomechanisms. This mechanism is a reflex arc, whose initial component is a receptor that measures pressure within the arterial system.

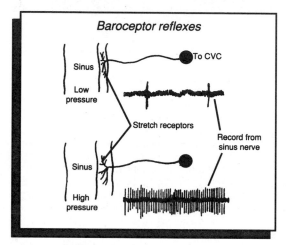

Baroceptor reflexes

Such sensors exist in the carotid sinus on the left and right sides, and the arch of the aorta. They consist of nerve fibres which ramify within the walls of the vessels concerned. If the pressure stretches the vessel, then the receptors initiate trains of action potentials. When blood pressure rises the frequency of discharge of impulses from the receptors increases. This has an inhibitory effect on the sympathetic outflow to both the arterioles and the heart. The arterioles are then dilated because the frequency of constrictor impulses is less (sympathetic tone is less). Consequently, peripheral resistance is decreased and this brings the blood pressure back on course again, i.e. lowers it. This is a negative feedback system and

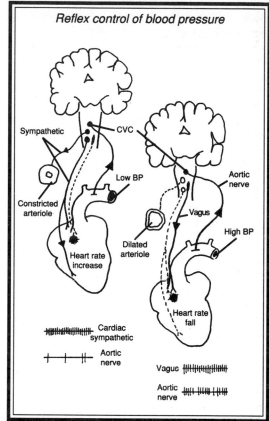

Reflex control of blood pressure

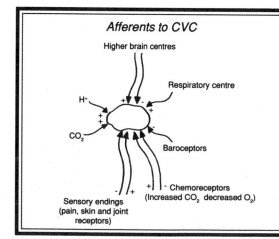

Afferents to CVC

There are also other afferent effects on the blood pressure. Many sensory receptors throughout the body send impulses to the cardiovascular centre which give rise to reflex changes in peripheral resistance, cardiac output, or both. These may be in conjunction with strenuous exercise where impulses arise in receptors that transmit to the cardiovascular centre. Another cause of altered cardiovascular parameters will be decreases in oxygen partial pressure or increases in carbon dioxide partial pressure in the blood. These stimulate chemoreceptors which exist in the same areas as the baroceptors. However, these mainly act upon the respiratory centre.

is the most important control of blood pressure. It is augmented by changes in the impulses going to the heart. In other words, if the blood pressure is above its normal values, heart rate will be slowed and this leads to a lowered cardiac output. So in most cases the response of the vasomotor centre and cardiac centre to unusual increases in blood pressure is a decrease in cardiac output and a decrease in peripheral resistance, both of which have negative feedback effects. In the reverse sense, the mechanism works if the pressure falls to too low a level, and the blood supply to the brain is threatened. The discharge of impulses by the receptors in the wall of the carotid sinus decreases, resulting in a reflex (homeostatic) augmentation of blood pressure.

Measurement of arterial pressure

The first measurement of blood pressure was made by an English clergyman in 1732, the Rev. Stephen Hales. He put a glass tube into the carotid artery of a mare and found that the height of the blood in the tube was about 8 feet. The level it reached was not constant; it went up and down with each heart beat, the maximum being just after systole, the minimum, diastole. The method for measuring blood pressure in experimental or research animals is roughly similar to this except that an artery is cannulated and a transducer measures the exact pressure within the vessel.

"In December I caused a mare to be tied down alive on her back....."

8.3 ft

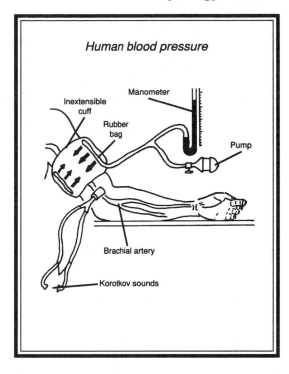

Human blood pressure

Inextensible cuff

Manometer

Rubber bag

Pump

Brachial artery

Korotkov sounds

Human blood pressure is measured indirectly using a blood pressure cuff wound around the arm. The cuff is made of inextensible cotton material within which is a rubber bladder. If inflated with a small handpump, this arm cuff can apply pressure to the arm and hence to the brachial artery. When the pressure in the cuff exceeds that in the artery, no blood can flow into the arm via the artery. As the pressure is gradually lowered within the cuff, blood can get under the cuff and into the arm once per heart beat. When it does this it makes a noise that can be detected with the stethoscope. These noises are called Korotkov sounds and are due to the blood coming through in spurts. When the sounds just appear the pressure is said to be systolic. As the pressure is lowered still further the sounds gradually disappear. When they just disappear the pressure is diastolic. Normal blood pressure in the human is 120/80, the top figure being systolic and the bottom figure being diastolic, both being in mmHg. In Western societies, arterial pressure rises with age and it is stated that normal systolic pressure is 100mmHg plus the age in years.

Relationship between venous pressure and cardiac output

The venous pressure (the venous return) determines the force of the heart beat. If the venous pressure and, hence, diastolic filling pressure rises, cardiac output is increased. This is because of the intrinsic relationship between length and tension in cardiac muscle fibres which determines that the force of contraction is proportional to their initial length.

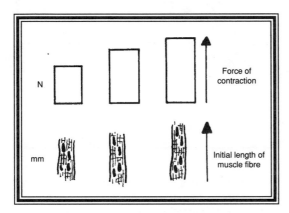

The venous return

When the blood has passed through the capillaries, the driving pressure has all but disappeared; it is 5 or 10 mmHg above atmospheric. Further on, the venous pressure is below atmospheric. However, such small pressures are sufficient to return the total cardiac output to the heart, because in the

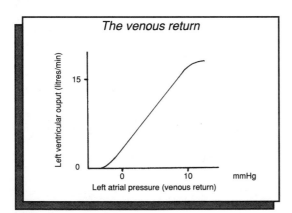

veins the resistance to flow is small. If the pressure rises in the right side of the heart (such as might result from leaking A-V valves) the pressure build up reduces the venous return. But, since the pressure is raised, the atrial muscle and later ventricle muscle fibres are stretched and force of the heartbeat is increased, therefore more blood is expelled by the ventricles. This is an autoregulatory mechanism which takes care of any discrepancy between venous return and cardiac output. It does so without the operation of any part of the nervous system.

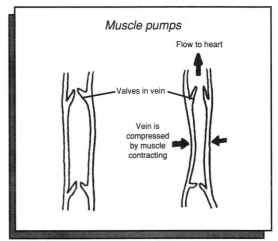

The venous return is also helped by the muscles acting as pumps. When a muscle contracts, it presses on the veins which have valves within them. Backflow is prevented by these valves and so the blood is driven towards the heart. The muscle pumps are more effective during exercise when an increase in venous return is in fact required.

Pulmonary circulation

The pulmonary circulation is in series with the systemic circulation and the heart pumps precisely the same amount of blood into the lungs as it does into the aorta. The pulmonary system, however, is much simpler because only one organ, the lung, has to be supplied. Therefore, there is little control of lung blood vessels. Also the resistance to flow, which depends on the pulmonary blood vessels, is low, and only low pressures are required. Normal pulmonary arterial pressure is 25/10mmHg (compared with 120/80 for the systemic system). The control of lung blood flow is anomalous. In muscle, for example, if there is a local deficiency of oxygen the arterioles are dilated in that region. In the lung, however, the effect is the reverse of this. When part of the lung is not ventilated sufficiently, it is of little use. Constriction takes place in the arterioles supplying the capillaries of this region and blood is shunted away from it.

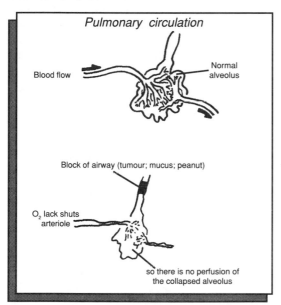

Pulmonary circulation

Blood flow

Normal alveolus

Block of airway (tumour; mucus; peanut)

O_2 lack shuts arteriole

so there is no perfusion of the collapsed alveolus

Circulation during exercise

During exercise, particularly if strenuous, demands are generated and changes take place that tend to upset the normal environmental parameters. Heat is produced, which may change body temperature. Carbon dioxide and waste products appear in much larger amounts than normal, the oxygen demand by active tissues is greatly increased and creates oxygen deficiency. The respiratory and circulatory systems are designed to respond to such situations.

The respiratory system ensures complete oxygenation of blood, and the circulatory system circulates the blood faster and changes the blood distribution. During exercise inputs reach the cardiovascular centre which mostly increase the frequency of the outflowing impulses in the sympathetic innervation to the heart and blood vessels.

Blood flow returning to the heart is increased as a result of the muscle pump action on the veins and the increased effect of negative pressure in the chest caused by deep breathing. The veins constrict under the influence of the sympathetic system; this reduces the volume of the circulatory bed and also adds to the increase in venous return. The extra volume of blood now entering the ventricles stretches cardiac muscle and, according to Starling's law of the heart, this gives rise to stronger contractions. The sympathetic nervous system increases both the force of contraction and heart rate, and, therefore, cardiac output. In severe exercise adrenaline may be secreted by the adrenal medulla and this will directly stimulate the SA node.

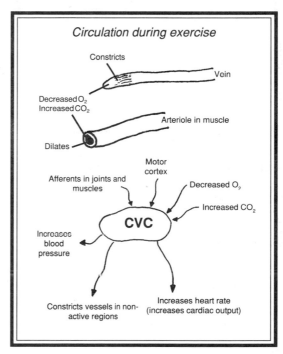

Circulation during exercise

Constricts

Vein

Decreased O_2
Increased CO_2

Arteriole in muscle

Dilates

Motor cortex

Afferents in joints and muscles

Decreased O_2

Increased CO_2

CVC

Increases blood pressure

Constricts vessels in non-active regions

Increases heart rate (increases cardiac output)

These factors will then raise the cardiac output up to 5 times the normal. At the same time, the cardiovascular system is diverting blood from those regions, such as the intestines, that are inactive, to the active ones, i.e. the muscles. Local dilation of arterioles in skeletal muscle and cardiac muscle takes place as a result of the production of metabolic products in these regions. The net effect is to raise muscle blood flow 20 or 30 times. As exercise proceeds, a large amount of heat is generated and this has to be removed from the body otherwise temperature would rise. This happens by a great increase in skin blood flow, so that heat may be lost by conduction and convection. Simultaneously, sweat glands are stimulated and heat is lost by the evaporation of sweat from the skin.

Regional blood flow in exercise (ml/min)

Region	At rest	Light exercise	Moderate exercise	Maximal exercise
Muscle	1000	4600	12400	22000
Heart	250	350	750	1000
Skin	400	1100	2000	600
Kidney	1100	900	600	250
Viscera	1000	900	600	300
Brain	750	750	750	750
Various (bone etc)	500	400	400	100
Total	5,000	9,000	17,500	25,000

CAPILLARY EXCHANGES AND THE LYMPHATIC SYSTEM

Capillaries

All the activities of the circulatory system are directed towards sending blood through the capillar-

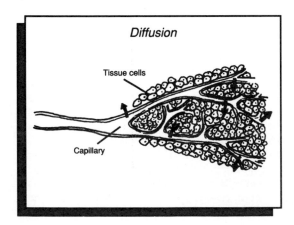

Diffusion

Tissue cells

Capillary

ies. The exchange of substances across the capillary cells to and from the tissue fluid, is mainly by diffusion. Diffusion only acts over short distances. Thus active cells such as muscle fibres need to have a profuse capillary blood supply, bone or fibrous tissue less so.

The capillary network is controlled on its input side by an arteriole which has a thick muscular wall and can be constricted to close down the blood supply. The arteriole acts as a tap. The degree of contraction in the muscle of this arteriolar tap is determined partly by local conditions in the muscle, and partly

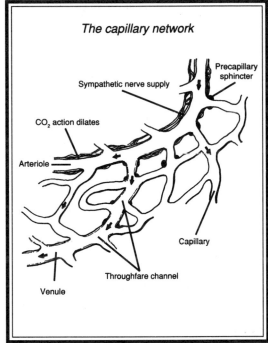

The capillary network

Precapillary sphincter

Sympathetic nerve supply

CO_2 action dilates

Arteriole

Capillary

Throughfare channel

Venule

by regulation of the sympathetic nervous supply. There are pre-capillary sphincters which are at the entrance to many capillaries. These control the blood flowing locally to different parts of a tissue.

Tissue fluid exchange

Two main mechanisms exist for transfer of substances between blood and tissue: diffusion and bulk flow. At the cellular level, diffusion is the more important.

Diffusion

Water with most dissolved materials will readily diffuse through the walls of capillaries. A substance that is at a higher concentration in the blood than it is in tissue fluid will leave the capillary by diffusion. Examples are oxygen and glucose. These molecules move along their concentration gradients. Waste metabolites and carbon dioxide will diffuse in the opposite direction, down their concentration gradients. Small amounts of protein travel from plasma into tissue fluid. Some of this is leakage; some proteins actually cross capillary walls enclosed in small vesicles (pinocytosis).

Cross section of a capillary

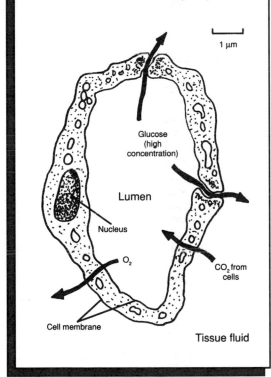

Bulk flow

Some fluid moves in and out of capillaries due to pressure gradients. This bulk flow is of water and all the substances dissolved in it, apart from proteins which are too large to pass through the pores.

The bulk flow is driven by, first of all, the blood pressure in the capillaries which forces fluid into tissue spaces. This process is known as filtration. It is opposed by osmotic pressure due to the plasma proteins. This causes fluid to return from tissues into capillaries. This process is known as reabsorption. Only proteins which can not diffuse through capillary walls exert osmotic effects.

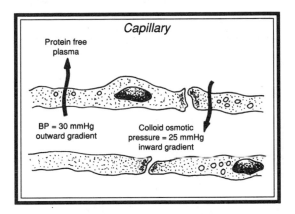

The net filtration at any point in a tissue is the sum of these forces. If the filtration exactly balances the reabsorption, then there will be no net fluid exchange in the tissue. Usually one part of the tissue, perhaps that nearest to the arterial end of the capillary network where pressures are highest, shows an excess of filtration over reabsorption, so the net fluid flow here is out from the capillaries into tissue spaces. Another part of tissue close by, probably near the venous end of the capillaries, has a lower hydrostatic pressure and the predominant process is one of reabsorption. At all events, if the tissue concerned is not to swell up or to become dehydrated, these two facets of exchange of water and solids across capillary membranes must balance.

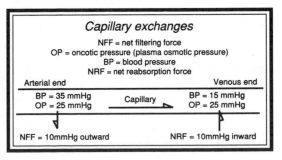

Under normal conditions, the total filtration is slightly more than the total reabsorption. The extra tissue fluid is collected by lymphatics, and returned via the lymphatic duct to the blood stream. If there is raised capillary pressure, or if there is a reduced protein osmotic pressure of the plasma, filtration will exceed the absorption. If the amount of tissue fluid is excessive, the capacity of the lymphatic system in removing it will be insufficient. Then fluid collects and the tissue spaces are filled. This is termed oedema.

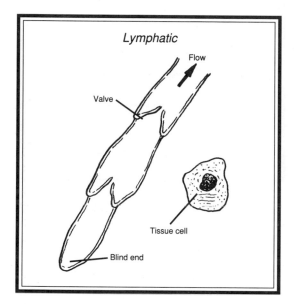

In the lungs, the blood pressure is low and does not usually exceed plasma protein osmotic pressure. Thus filtration does not occur.

Lymphatic system

Lymphatics are blind ending, small vessels like venules, complete with one-way valves, so that the drainage of lymph is from the periphery to the centre. They return protein and excess tissue fluid that results from the filtration and reabsorption processes to the circulatory system. The blind-ended lymphatic capillaries then join to form thin walled lymphatic vessels. These run to the regional lymph nodes, eventually to empty via two main lymphatic ducts into the great veins in the neck. The structure of lymphatics is slightly different from that of a blood capillary, being more porous

and having an arrangement of cells overlapping so that fluid can flow into them easily, but flowing out is difficult. Most of the wall of the lymphatics, in other words, is one large valve. When pressure oscillates within a tissue, such as it does during contraction of muscle, the lymphatics will act as a pump in roughly the same way as the veins. However, the total flow of lymph is very much less than the venous return from any limb or organ, being only about 5 litres of lymph per day entering the blood stream.

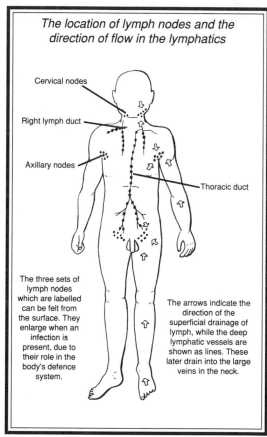

The lymphatic system has other functions apart from that of returning tissue fluid to the blood stream. It is the route of fat absorption from the small intestine and after a fatty meal the lymph becomes milky. It also provides a means of transport for lymphocytes and hence has an important role in the immune system (see Chapter 4).

Filtration in the lymph nodes

Particulate matter is removed from the lymph as it passes through the lymph nodes (or lymph "glands"). Lymph reaches the node through the afferent lymphatics round its periphery. It leaves via the efferent lymphatics at the hilus. The circulation inside the node is through the medullary sinuses. These sinuses are lined with phagocytes which remove particulate matter such as dead cells and bacteria from the lymph. The lymph nodes are also very important organs of the immune system, since they are able to produce specific lymphocytes and plasma cells in response to antigens (see Chapter 4).

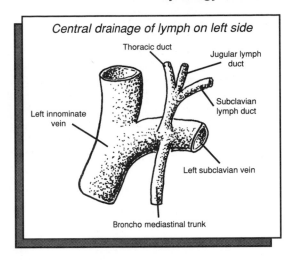

Central drainage of lymph on left side

Thoracic duct

Jugular lymph duct

Subclavian lymph duct

Left innominate vein

Left subclavian vein

Broncho mediastinal trunk

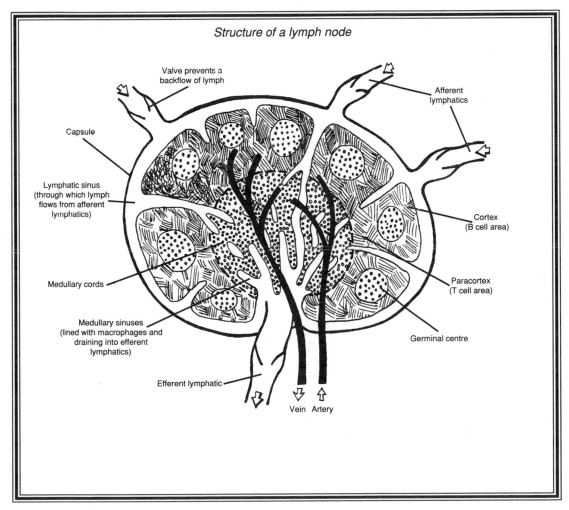

Structure of a lymph node

Valve prevents a backflow of lymph

Afferent lymphatics

Capsule

Lymphatic sinus (through which lymph flows from afferent lymphatics)

Cortex (B cell area)

Medullary cords

Paracortex (T cell area)

Medullary sinuses (lined with macrophages and draining into efferent lymphatics)

Germinal centre

Efferent lymphatic

Vein   Artery

MUSCLE

MUSCLE AND CONTRACTION

Muscle is a tissue that shortens and develops tension so that movement is brought about. There are three different kinds of muscle associated with different functions. Skeletal muscle moves the skeleton; cardiac muscle is in the heart and smooth muscle is concerned with movements of the visceral organs. Skeletal muscle and cardiac muscle fibres under the microscope appear to be striated whilst smooth muscle does not. Skeletal muscle is under the control of the nervous system and can be contracted or relaxed at will; it is therefore called voluntary muscle. Although cardiac muscle and smooth muscle are also under nervous control, this is much less direct and most persons are unable to achieve any degree of voluntary control over them.

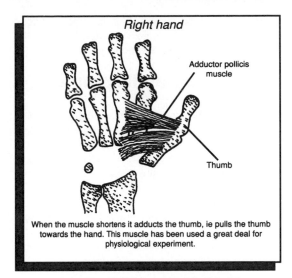

Right hand

Adductor pollicis muscle

Thumb

When the muscle shortens it adducts the thumb, ie pulls the thumb towards the hand. This muscle has been used a great deal for physiological experiment.

SKELETAL MUSCLE

Most muscles which are attached to the skeleton are made up of many long thin cells called muscle fibres that are in parallel. They are contained in connective tissue sheets and form bundles of fibres.

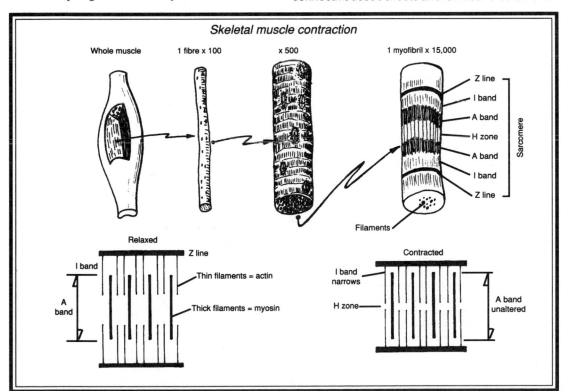

Skeletal muscle contraction

Whole muscle | 1 fibre x 100 | x 500 | 1 myofibril x 15,000

Z line
I band
A band
H zone
A band
I band
Z line

Sarcomere

Filaments

Relaxed

I band

A band

Z line

Thin filaments = actin

Thick filaments = myosin

Contracted

I band narrows

H zone

A band unaltered

These are then bound in further connective tissue to form the whole muscle. At each end are the tendons of origin and insertion.

The contractile process

Each muscle fibre consists of many myofibrils. The contractile mechanism exists at the microscopic level in each myofibril. A contraction of the whole muscle is produced by large numbers of myofibrils undergoing the contractile process. This has three parts to it:-

• Excitation; the action potential spreading over the membrane of the muscle fibre.

• The contraction of the muscle fibre.

• A system which links excitation and contraction known as excitation-contraction coupling.

Excitation

In order for the contractile system in a muscle fibre to be activated it is necessary for its membrane to be depolarised. An action potential (a depolarisation) spreading throughout the membrane of the muscle fibre normally results from a nerve impulse arriving at the neuromuscular junction. The muscle fibre membrane adjacent to the motor nerve terminal is known as the end-plate region. When the nerve action potential arrives in the terminal, acetylcholine is released and the end-plate region of the membrane undergoes local depolarisation. This generates an action potential in the rest of the membrane of the muscle fibre which propagates to the ends of the fibre.

Strength-duration relations

In any excitable tissue the strength of a stimulus required to produce a threshold response depends on the duration of the stimulus. The form of the curve relating strength and duration is hyperbolic. Curves for motor nerve fibres and skeletal muscle fibres are different. As far as muscle is concerned it is possible to stimulate fibres directly and this requires stronger and longer-lasting electrical pulses than are required to stimulate the nerve fibre.

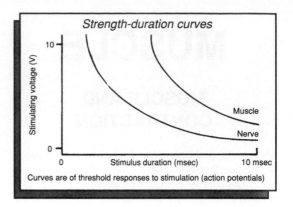

Curves are of threshold responses to stimulation (action potentials)

Strength response curve

If a single muscle fibre is directly stimulated, the curve of force looks like (a). The threshold strength and the maximal stimulus strength are identical. The fibre obeys the "all or nothing" law. For the whole muscle containing a large number of fibres the relationship between stimulus strength and muscle response is S-shaped. The threshold stimulus strength and the maximal stimulus strength

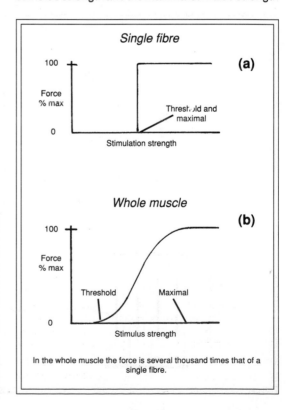

In the whole muscle the force is several thousand times that of a single fibre.

differ (b). The relationship is this shape because any muscle contains a large number of fibres, each of which has a threshold stimulus strength slightly different from that of all the others. The threshold for the whole muscle is that of the threshold of the most sensitive fibres; the maximal stimulus strength for the whole muscle is that required to activate the least sensitive fibres.

Theoretically, the curve (b) should be made up of a large number of individual steps, each step being when one muscle fibre is activated. When the muscle is stimulated through its motor nerve, the steps indeed may be apparent because there are fewer of them. This is due to the fact that we are now considering the effect of a number of motor units, each motor unit consisting of, perhaps, one hundred separate fibres, so the steps will be a hundred times larger.

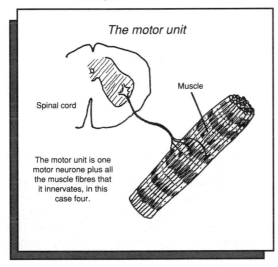

The motor unit

Spinal cord

Muscle

The motor unit is one motor neurone plus all the muscle fibres that it innervates, in this case four.

The sliding filament model

If you look at a muscle under the light microscope you find that it consists of a large number of muscle fibres. In each muscle fibre is a number of myofibrils. These have a characteristic striated appearance. If you now examine the muscle fibre under the electron microscope, myofibrils can be seen each to consist of a series of interdigitating thin and thick filaments. The thin filaments are in a square latticework attached to a dense structure called the Z-line.

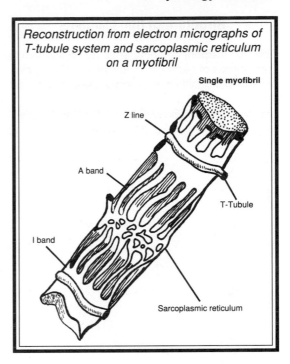

Reconstruction from electron micrographs of T-tubule system and sarcoplasmic reticulum on a myofibril

Single myofibril

Z line

A band

T-Tubule

I band

Sarcoplasmic reticulum

The thin filaments are fixed into the Z-line and they continue on through the I-band and into the A-band. The A-band is highly refractile and corresponds to the location of the thick filaments. The H-zone is a region of thick filaments where no overlap occurs between thick and thin filaments. If you now look end-ways on at a myofibril you find that the thick filaments are in a hexagonal pattern.

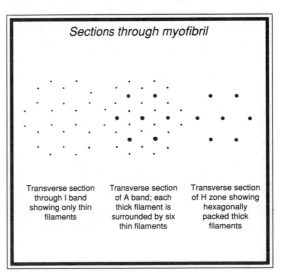

Sections through myofibril

Transverse section through I band showing only thin filaments

Transverse section of A band; each thick filament is surrounded by six thin filaments

Transverse section of H zone showing hexagonally packed thick filaments

Structure of thin filaments

The thin filament contains actin. Each filament is two strands of f-actin, in a right handed helix, plus additional proteins which control the contraction.

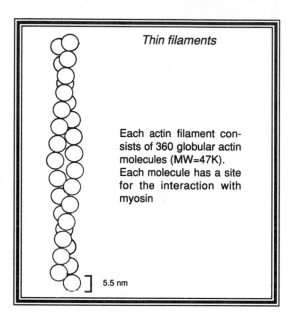

Thin filaments

Each actin filament consists of 360 globular actin molecules (MW=47K). Each molecule has a site for the interaction with myosin

5.5 nm

Structure of thick filaments

Thick filaments are composed of myosin, a high molecular weight protein, which appears in the electron microscope as a rod with a globular head. It is both part of muscle structure and also an enzyme. The sliding filament theory was put forward by A. F. Huxley and independently by H. E. Huxley (no relation) in 1954. The main observation

was that the A-band (myosin) does not change in length if the muscle contracts or is stretched. This must mean that the thick filaments slide past the thin filaments when the muscle contracts or relaxes. There are cross bridges between the thick and the thin filaments where these overlap. These cross-bridges act like little legs which propel the thin filaments past the thick ones.

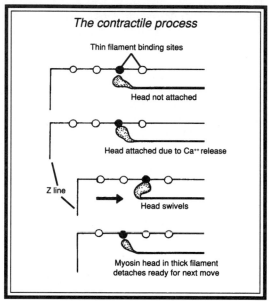

The contractile process

Thin filament binding sites

Head not attached

Head attached due to Ca++ release

Z line

Head swivels

Myosin head in thick filament detaches ready for next move

The head of each myosin cross bridge has on It an actin binding site and an ATPase. When ATP is hydrolysed, energy is released, movement of the cross-bridge occurs and sliding takes place. The way in which the energy released from ATP produces the sliding is unknown at the moment.

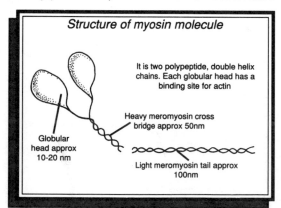

Structure of myosin molecule

It is two polypeptide, double helix chains. Each globular head has a binding site for actin

Heavy meromyosin cross bridge approx 50nm

Globular head approx 10-20 nm

Light meromyosin tail approx 100nm

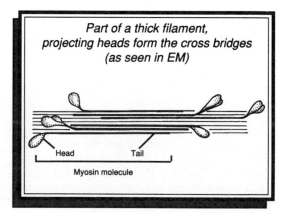

Part of a thick filament, projecting heads form the cross bridges (as seen in EM)

Head   Tail

Myosin molecule

Isometric contraction

Skeletal muscles can be excited either directly (a stimulating current is applied to the muscle fibre) or via its motor nerve. The force generated when a muscle contracts is known as its tension and is measured in Newtons per square metre of cross

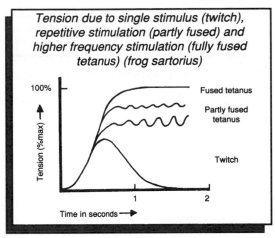

Tension due to single stimulus (twitch), repetitive stimulation (partly fused) and higher frequency stimulation (fully fused tetanus) (frog sartorius)

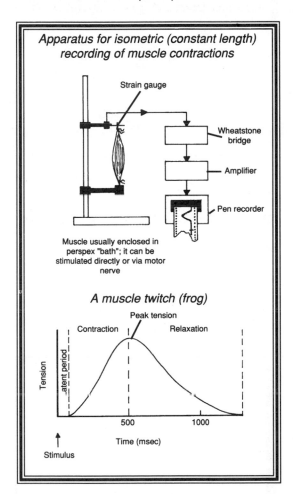

Apparatus for isometric (constant length) recording of muscle contractions

Strain gauge

Wheatstone bridge

Amplifier

Pen recorder

Muscle usually enclosed in perspex "bath"; it can be stimulated directly or via motor nerve

A muscle twitch (frog)

Peak tension

Contraction   Relaxation

Latent period

Tension

Stimulus

500   1000

Time (msec)

is then called a tetanus. A tetanus may be unfused (shocks are spaced far enough apart for the tension to decay in between each), or it may be fused (a smooth contraction). When twitches are repeated in this way to give rise to a tetanus, the elastic elements within the muscle get stretched and the full tension due to that muscle is produced.

Length-tension relationships

Muscle is quite an elastic structure because it contains elastic fibres. These are arranged either in series with the muscle (series elastic component) or in parallel. If one measures the tension produced by a muscle it is a compound of active tension produced during the contraction by the contractile process and passive tension due to the elastic components. It is a general rule that the maximum tension produced by a muscle occurs at normal muscle length in the body.

section. The greatest tension exerted by muscles in the mammal is around $5 \times 10^5 \text{N.m}^{-2}$. A muscle twitch lasts for only a short length of time and the tension then falls. When a second shock is given before the tension drops back to zero, the second response of the muscle adds on to the first one. The tension is larger when you give two shocks within a small time interval than when one alone is given. If a train of shocks is given, the contraction is maintained and

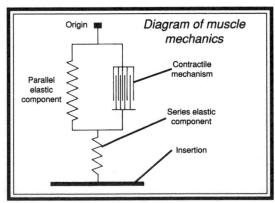

Diagram of muscle mechanics

Origin

Contractile mechanism

Parallel elastic component

Series elastic component

Insertion

Experiments show that at lengths shorter than in the body, tension produced gets less, and likewise

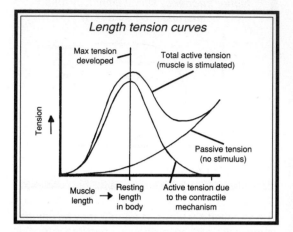

Length tension curves

at longer lengths. The reason for this is that the sliding filaments overlap to a maximum when the muscle is at body length. If you stretch it beyond this point, fewer cross-bridges will be engaged; tension will then be less. Conversely, if the muscle is at a shorter length, the sliding filaments will not be able to slide any further because they will come to a full stop when they collide with the Z-band.

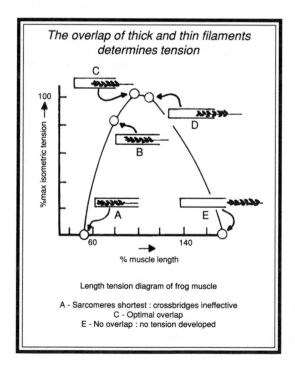

The overlap of thick and thin filaments determines tension

Length tension diagram of frog muscle

A - Sarcomeres shortest : crossbridges ineffective
C - Optimal overlap
E - No overlap : no tension developed

Isotonic contractions

An isotonic contraction occurs when the muscle is allowed to shorten as it is stimulated. The less the

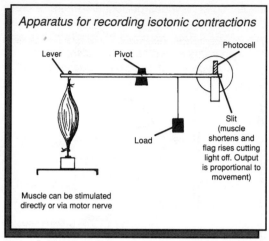

Apparatus for recording isotonic contractions

Muscle can be stimulated directly or via motor nerve

load on the muscle the faster the muscle can contract. This relationship between the velocity of shortening and the force developed (tension) is called the force-velocity relation. The precise shape of this curve is supposed to be related to mechanical restraints imposed on the splitting of ATP by cross-bridges.

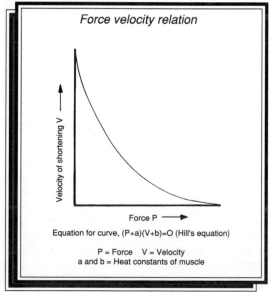

Force velocity relation

Equation for curve, $(P+a)(V+b) = O$ (Hill's equation)

P = Force   V = Velocity
a and b = Heat constants of muscle

Control of contraction

If one takes a thin filament by itself it will react with myosin and develop tension even if there is no calcium present. Normally, however, regulatory proteins are attached to the actin in the thin filaments and this prevents any reaction with myosin, until the ionised calcium concentration exceeds a critical level. These regulatory proteins are called tropomyosin and troponin. Their presence on the actin enables the whole system to be controlled by varying the ionised calcium concentration. When muscle is relaxed, the calcium concentration is low, tropomyosin gets in the way of the myosin combining sites on the actin, whereupon the activity of cross-bridges is prevented. When an action potential reaches a muscle, calcium is released from the vesicles in the sarcoplasmic reticulum and binds with troponin, resulting in conformational changes and the myosin combining site is uncovered. Now cross-bridges can form and tension is thus produced.

Calcium and contraction

When the muscle fibre contracts there is a rapid large increase in ATP splitting by the cross-bridges. This is due to large amounts of calcium appearing in the cytoplasm when the excitation-contraction cycle is initiated. At rest, calcium concentration outside the myofibrils is about 10^{-7}mol.l^{-1}. With this concentration, cross-bridge activity does not occur. This low calcium concentration is due to calcium being stored in the sarcoplasmic reticulum, which is a complex system of membrane bound cavities within the muscle. The calcium in the sarcoplasmic reticulum is combined with calcium binding proteins. The internal membrane system of the sarcoplasmic reticulum is a network of vesicular elements around the myofibrils and is in apposition to the transverse tubular system (T system) connected to the extra-cellular space. When an action potential is propagated along the muscle fibre membrane, it is propagated down the transverse tubules. From these it is passed to the sarcoplasmic reticulum and calcium is released into the spaces surrounding the myofilaments. This released calcium activates the cross-bridges and contraction takes place. During activity, calcium levels rise to around 10^{-5}mol.l^{-1}. When contraction is over, the calcium is pumped back into the sarcoplasmic reticulum by the sarcoplasmic reticulum ATPase.

Energy source for contraction

The only immediate energy source for the contraction of muscle is ATP. This ATP must be continually renewed. In muscle cells ATP is produced during exercise, mainly because of oxidation of the stores of glycogen. When strong contractions are being made this process may be inadequate. There are three further processes which can take place.

• Oxygen supply

Glycogen requires oxygen for its oxidation. Muscles contain a red-pink pigment, myoglobin which is similar to haemoglobin. Myoglobin forms complexes with oxygen and acts as a store. When the concentration of oxygen in the cell is low, myoglobin releases its oxygen.

• Anaerobic metabolism

In addition to getting ATP from the complete oxidation of glycogen, a muscle cell can break down further glycogen part of the way, with the formation of lactic acid. This reaction does not need oxygen and is called anaerobic metabolism.

• Storage of high energy phosphate

In a single muscle fibre there is comparatively little ATP storage. It can, however, be regenerated quickly from high energy creatine phosphate which is stored in larger amounts.

ADP + creatine phosphate ⟶ ATP + creatine

This reaction takes place when ATP levels are low. It can continue until all the creatine phosphate is used up. It is later restored by means of the converse reaction during muscle relaxation.

The efficiency of a muscle when it contracts is roughly 25%. In other words, in an isotonic contraction about a quarter of the energy expended by the muscle is used to do work. The remaining three-quarters is degraded as heat.

During an isometric contraction there is no external work done and it could be said that 100% of the energy expended disappears as heat.

Types of skeletal muscle fibre

Generally, there are three types of muscle fibre in mammals. They differ in their ability to generate ATP and also in their speed of contraction and resistance to fatigue. Their properties are summarised in the table.

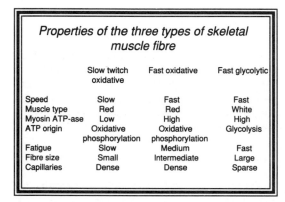

Properties of the three types of skeletal muscle fibre			
	Slow twitch oxidative	Fast oxidative	Fast glycolytic
Speed	Slow	Fast	Fast
Muscle type	Red	Red	White
Myosin ATP-ase	Low	High	High
ATP origin	Oxidative phosphorylation	Oxidative phosphorylation	Glycolysis
Fatigue	Slow	Medium	Fast
Fibre size	Small	Intermediate	Large
Capillaries	Dense	Dense	Sparse

CARDIAC MUSCLE

Cardiac muscle resembles skeletal muscle. It has striations and also contains myofibrils with thick and thin filaments of myosin and actin. The underlying mechanisms of contraction are just the same in the two types of muscle.

The cardiac muscle cell is, however, a single cell, has a single nucleus and is joined to other cells with intercalated discs. The cells in cardiac muscle are also different from skeletal muscle in that there are gap junctions where the plasma membrane of two cells is fused together. At gap junctions, action potentials can spread from one cell to another. Cardiac muscle cells contain many mitochondria, have a very full blood supply and only a moderately developed sarcoplasmic reticulum. The active state and also the action potential in cardiac muscle last much longer than in skeletal muscle. For example, a typical cardiac muscle action potential will last for about 300msec and the refractory period is of the same duration. These facts imply that it is not possible to produce a tetanus in cardiac muscle.

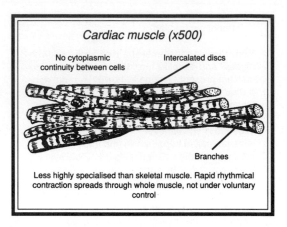

Cardiac muscle (x500)

No cytoplasmic continuity between cells

Intercalated discs

Branches

Less highly specialised than skeletal muscle. Rapid rhythmical contraction spreads through whole muscle, not under voluntary control

SMOOTH MUSCLE

The contractions in smooth muscle have a longer duration, they are more variable and they produce less tension than in skeletal muscle. Also, structurally, smooth muscle differs considerably from skeletal muscle. The cells are much smaller, spindle shaped, and contain one nucleus only. There are no cross striations because the actin and myosin filaments are not divided into sarcomeres. Two main types of smooth muscle are found in the body; visceral and multiunit.

Visceral smooth muscle

This is found mainly in the internal organs, especially in the urinary tract, reproductive system and digestive tract.

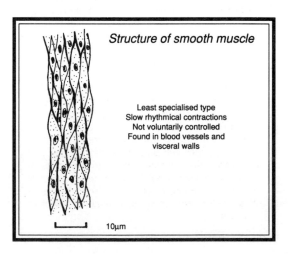

Structure of smooth muscle

Least specialised type
Slow rhythmical contractions
Not voluntarily controlled
Found in blood vessels and visceral walls

10µm

The smooth muscle cells form sheets and tubes, the cell membranes sticking to one another. In consequence, an action potential can spread from cell to cell. The contractions occurring in visceral smooth muscle spread in waves from any stimulation point. An example of this is seen in the peristaltic waves that occur in the intestines.

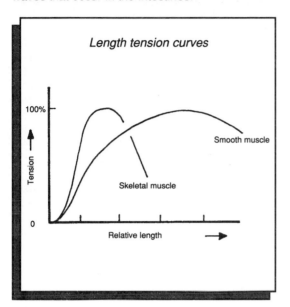

Length tension curves

Characteristics of skeletal, heart and smooth muscle		
Skeletal	Heart	Smooth
Fixed to bony origins and insertions	It is the muscle of heart not attached to bone	Found in blood vessels and viscera
Multinucleate cells composed of many parallel fibres. Cells are long and cylindrical often stretching whole length of muscle	Cells have single nuclei and form a branching network	Cells are spindle shaped each with one nucleus
Cross-striations	Cross-striations present	Cross-striations not clear
Contraction fast	Contraction slow	Contraction very slow
Action potential is confined to one fibre and does not spread	Action potential spreads throughout ventricles or atria	Action potential travels from one cell to another
Only contracts when motor neurone is activated	Contracts ryhthmically and spontaneously	Rhythmical contractions
Innervated by somatic nerves	Innervated by autonomic nerves	Innervated by autonomic nerves

Multiunit smooth muscle

Smooth muscle, comprised of many units, is found in bodily sites where finely graded movements are required. The properties of multiunit smooth muscles are between those of skeletal and visceral smooth muscle. They consist of separate motor units responding only to the activity of motor nervoc.

Another characteristic of smooth muscle is its tendency to give rise to spontaneous contractions. These occur even in isolated smooth muscle when there is no external nerve supply. Also, visceral smooth muscle responds to stretch by contracting. It is plastic in the sense that it can have very different resting lengths in spite of the tension being the same. When urine distends the bladder, the smooth muscle is stretched and it develops more tension. However, the tension soon decreases, and the volume is increased without any great rise in its internal pressure.

These spontaneous contractions of smooth muscle are, to some extent, controlled in the body by motor neurones which innervate the muscle and by a number of chemical substances including hormones.

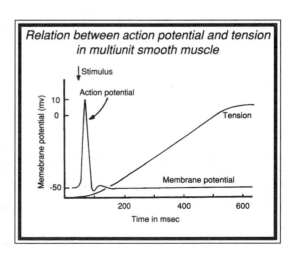

Relation between action potential and tension in multiunit smooth muscle

Action potentials do not travel from one cell to another. Neither do multiunit smooth muscles exhibit spontaneous contractions. Multiunit smooth muscle is able to contract like a twitch in response to a single stimulus. Nevertheless, this twitch contraction is a good deal slower than that seen in skeletal muscle; it is not under voluntary control.

NEUROMUSCULAR JUNCTION

The junction between a motor neurone and a skeletal muscle fibre is called the neuromuscular junction. The axon ends in contact with a special region of membrane on the surface of the muscle fibre known as the end-plate. The axon terminal and the end-plate directly beneath it comprise the neuromuscular junction. Each action potential arriving in the motor nerve always produces a muscle action potential, which then spreads along the membrane of the muscle fibre giving rise to a contraction of that fibre.

Each myelinated nerve fibre arriving at the junction has a number of terminal non-myelinated branches, each about 1µm in diameter running in grooves on the surface membrane of the muscle. For the length of this groove, the synaptic contact occurs between nerve and muscle membrane. The surface area of this region is increased by many folds in the membrane.

If one records electrically with a microelectrode from such a synapse it is found that a delay of about 0.5msec occurs between the nerve impulse reaching the terminal and the action potential being produced in the muscle membrane. It has been found that the delay is a result of time taken for the release of acetylcholine, and its diffusion from the presynaptic membrane across the synaptic cleft and its reaching the postsynaptic membrane.

The evidence for the existence of these processes is that:-

• Acetylcholine is released from the presynaptic

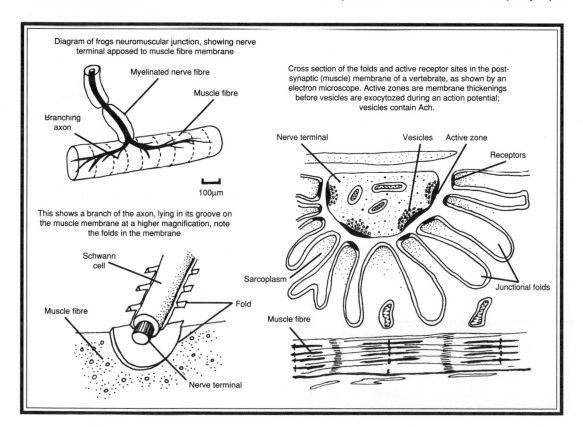

Diagram of frogs neuromuscular junction, showing nerve terminal apposed to muscle fibre membrane

Myelinated nerve fibre

Muscle fibre

Branching axon

100µm

This shows a branch of the axon, lying in its groove on the muscle membrane at a higher magnification, note the folds in the membrane

Schwann cell

Muscle fibre

Fold

Nerve terminal

Cross section of the folds and active receptor sites in the post-synaptic (muscle) membrane of a vertebrate, as shown by an electron microscope. Active zones are membrane thickenings before vesicles are exocytozed during an action potential; vesicles contain Ach.

Nerve terminal

Vesicles

Active zone

Receptors

Sarcoplasm

Junctional folds

Muscle fibre

terminal membrane.

• Acetylcholine, if artificially introduced into the cleft, will produce an action potential in the postsynaptic membrane.

• Enough acetylcholine is normally released by an action potential to produce another in the muscle fibre.

In the presynaptic region of the neuromuscular junction, the acetylcholine is stored in the cytoplasm in vesicles. When a nerve impulse arrives, the membrane of the nerve terminal is depolarised, calcium ions enter the cell from the outside, produce fusion between the vesicles and the nerve membrane. The vesicles then release the acetylcholine into the synaptic cleft. When the muscle membrane is reached by the acetylcholine molecules, depolarisation is produced; this is called the end-plate potential . Immediately, the acetylcholine left within the cleft, is hydrolysed to choline by an enzyme, acetylcholinesterase. The choline that is produced in this way gets absorbed again into the presynaptic nerve terminal and is utilised to produce more acetylcholine.

The vesicles are randomly released at all neuromuscular junctions, roughly once or twice every second. The amount of acetylcholine released from a vesicle is called a quantum. Each quantum gives a tiny depolarisation of a muscle cell called a miniature end-plate potential (MEPP). Notice that when recording from the end-plate region it is found that the end-plate potential is of the order of 75mV, whereas the miniatures are of the order of 1mV each. In addition, end-plate potentials only occur near to the neuromuscular junction; they do not spread. They are due to opening ionic channels which do not discriminate between small cations. The action of acetylcholine is to render the membrane permeable to sodium ions, potassium ions and calcium ions. When the endplate potential reaches a critical threshold level it initiates an action potential in the muscle membrane.

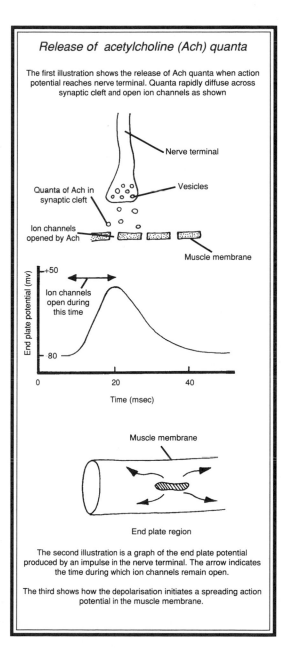

Release of acetylcholine (Ach) quanta

The first illustration shows the release of Ach quanta when action potential reaches nerve terminal. Quanta rapidly diffuse across synaptic cleft and open ion channels as shown

Nerve terminal

Quanta of Ach in synaptic cleft

Vesicles

Ion channels opened by Ach

Muscle membrane

Ion channels open during this time

End plate potential (mv)

Time (msec)

Muscle membrane

End plate region

The second illustration is a graph of the end plate potential produced by an impulse in the nerve terminal. The arrow indicates the time during which ion channels remain open.

The third shows how the depolarisation initiates a spreading action potential in the muscle membrane.

KIDNEY

The kidney regulates body fluids and excretes waste products. It is a rapid process because 1/5th of the blood volume goes through the two kidneys each minute.

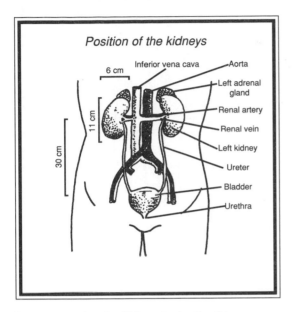

Position of the kidneys

In cross section the kidney looks like this:-

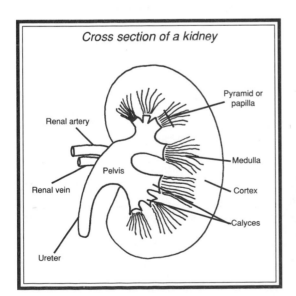

Cross section of a kidney

Each human kidney is made up of a million nephrons.

A NEPHRON

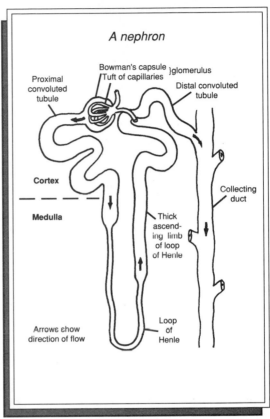

A nephron

Several distal convoluted tubules lead to a collecting duct. Many of these then join and empty into the renal calyx.

Dimensions of a nephron		
	length mm	diameter µm
Proximal convoluted tubule	12 - 24	50 - 65
Thin limbs of loop of Henle	0 -14	14 - 22
Distal convoluted tubule	2 - 9	20 - 50
Total length of nephron	20 - 44	
Glomerulus		150 - 250

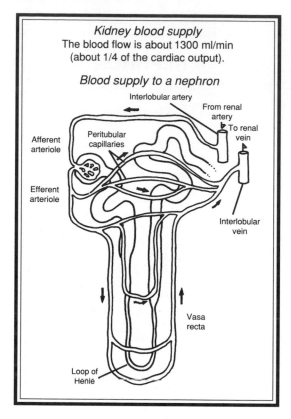

Kidney blood supply
The blood flow is about 1300 ml/min
(about 1/4 of the cardiac output).

Blood supply to a nephron

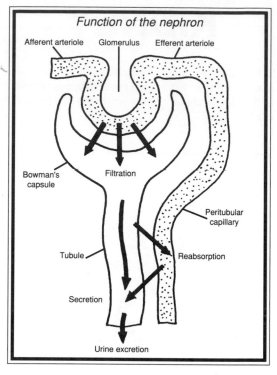

Function of the nephron

Glomerular filtration

The glomerular capillaries filter the blood. The large volume of fluid thus produced in Bowman's capsule is called glomerular filtrate. It contains no solid elements and very few plasma proteins.

The blood constituents of less than 68,000 MW can pass into the tubule, but anything larger is retained in the blood.

The structural basis for the sieve-like properties of the glomerulus is probably due to a gelatinous basement membrane. This separates the endothelial cells of the capillaries from the epithelial cells that line Bowman's capsule.

URINE FORMATION

Urine is formed from the blood by filtration from the glomerular capillaries into Bowman's capsule. This produces a large volume of protein-free plasma. The composition of the filtrate is then modified as it flows down the rest of the nephron.

Substances are easily transferred from tubules to capillaries (and vice versa) because they are close to each other. Absorption is from lumen to capillary: secretion is the opposite. Both vary over a wide range. Useful substances are completely reabsorbed while most waste products are excreted.

Normal urine

The daily output of urine varies widely: it is normally between 1 and 2 litres. At least 300 ml/day are needed for removal of the body's waste products. These are mostly urea, uric acid and creatinine.

• Urea is the end product of protein metabolism and is present in large amounts.

• Uric acid comes from nucleic acids and purines.

• Creatinine is a waste product of muscle metabolism: the quantity is related to muscle mass, being greater in a man who is muscular than in a child or woman.

Production rate	0.3 - 20 ml/min
Specific gravity	1.002 - 1.03
pH (depends on diet)	4.5 - 8.2

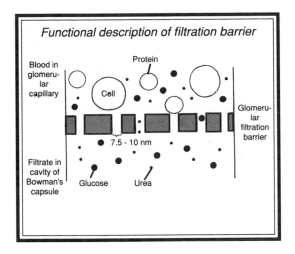

Functional description of filtration barrier

Factors altering GFR

- Concentration of plasma proteins.

- Intratubular hydrostatic pressure.

- Capillary blood pressure. This is probably the most important influence. Changes in diameter of either the afferent or efferent arteriole will affect the pressure in the glomerular capillaries. Moreover, if arterial pressure changes, the arterioles can largely compensate for this and keep the capillary pressure the same.

Consequently it is possible for the glomerular filtration rate to be maintained constant despite wide variations in systemic blood pressure.

Filtration pressure

Filtration will only occur if there is a large enough pressure to force the filtrate through the barrier. This filtration, under pressure, through an ultrafine filter is referred to as ultrafiltration.

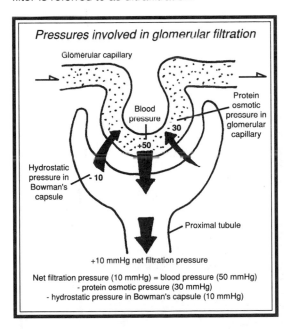

Pressures involved in glomerular filtration

Net filtration pressure (10 mmHg) = blood pressure (50 mmHg)
- protein osmotic pressure (30 mmHg)
- hydrostatic pressure in Bowman's capsule (10 mmHg)

The rate of fluid transfer from capillaries into Bowman's capsule is called the glomerular filtration rate (GFR): it is proportional to the effective filtration pressure.

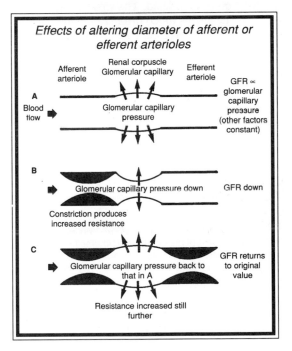

Effects of altering diameter of afferent or efferent arterioles

Control of capillary blood pressure and hence glomerular filtration rate by changes in arteriolar resistance is called "autoregulation". Autoregulation does not depend on any neural control of the renal arterioles. However it will fail if the systemic blood pressure falls below 50-60 mmHg, and this will result in anuria or absence of glomerular filtration.

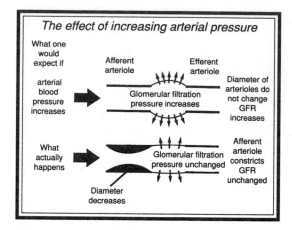

The effect of increasing arterial pressure

What one would expect if arterial blood pressure increases → Afferent arteriole / Efferent arteriole / Glomerular filtration pressure increases / Diameter of arterioles do not change GFR increases

What actually happens → Glomerular filtration pressure unchanged / Afferent arteriole constricts GFR unchanged / Diameter decreases

As the glomerular filtrate passes onto the other subdvisions of the nephron, its volume and composition are greatly modified. The volume of glomerular filtrate is always much greater than the volume of urine excreted over the same period of time.

The relationship between the rate of formation of glomerular filtrate and the rate of urine formation.

	Amount ml/min	= litre/24 hours
Cardiac output (CO)	5000	7200
Blood flow to kidneys (1/4 CO)	1250	1800
Formation of filtrate (1/10 blood flow)	125	180
Formation of urine	1	1.5
Reabsorption of filtrate	124	178.5

The glomerular filtration rate is defined as the volume of plasma filtered in unit time, = 180 litres/day in man. The total volume of plasma in man is approximately 3 litres. Therefore the entire volume of plasma is filtered 60 times each day.

TUBULAR FUNCTION

The kidney's ability to regulate the internal environment depends on it varying the reabsorption rates of some substances, and the secretion rates of others. Reabsorption and secretion occur in the tubules of the nephron. Tubular function should be considered with the following details in mind.

• The quantities of materials entering the nephron via the glomerular filtrate are enormous.

• Sizeable fractions of waste products are excreted.

• Only small fractions of the filtered "useful" plasma components (e.g. water, glucose, electrolytes) are excreted.

The table shows how some components are handled by filtration and reabsorption.

	Amount filtered	Amount excreted	%Reabsorbed
Water litres	180	1.8	99
Sodium g	630	3.2	99.5
Glocose g	180	0	100
Urea g	56	28	50

Transport processes involved in tubular reabsorption or secretion can be categorised broadly as active or passive (see Chapter 2). Transport of substances to and from the renal tubules involves crossing a sequence of membranes.

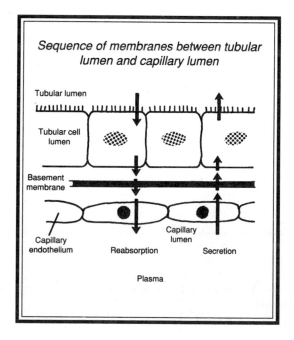

Sequence of membranes between tubular lumen and capillary lumen

Tubular lumen

Tubular cell lumen

Basement membrane

Capillary endothelium

Capillary lumen

Reabsorption

Secretion

Plasma

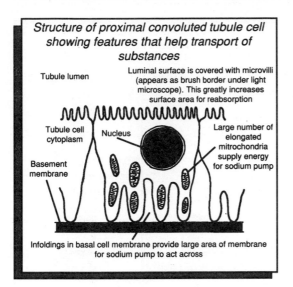

Structure of proximal convoluted tubule cell showing features that help transport of substances

Tubule lumen

Luminal surface is covered with microvilli (appears as brush border under light microscope). This greatly increases surface area for reabsorption

Tubule cell cytoplasm

Nucleus

Large number of elongated mitrochondria supply energy for sodium pump

Basement membrane

Infoldings in basal cell membrane provide large area of membrane for sodium pump to act across

Most of the sodium ion reabsorption occurs in the proximal tubules. Between 2/3rds and 7/8ths of the filtered sodium ions are reabsorbed there.

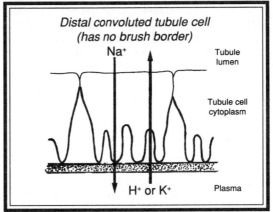

Distal convoluted tubule cell (has no brush border)

Na^+

Tubule lumen

Tubule cell cytoplasm

H^+ or K^+

Plasma

How the tubules reabsorb specific substances

Sodium

Reabsorption of sodium ions is quantitatively the most important active process in the kidney, and involves a sodium "pump". Transport of the sodium ions is accompanied by appropriate anions (principally chloride ions in the proximal tubules). There is also a concomitant flow of water which maintains the osmotic equilibrium.

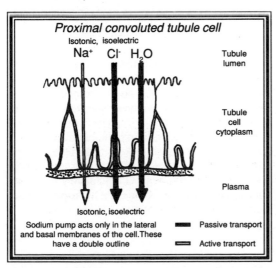

Proximal convoluted tubule cell

Isotonic, isoelectric

Na^+ Cl^- H_2O

Tubule lumen

Tubule cell cytoplasm

Plasma

Isotonic, isoelectric

Sodium pump acts only in the lateral and basal membranes of the cell. These have a double outline

▬ Passive transport

▭ Active transport

Glucose

Under normal conditions glucose is completely reabsorbed actively. However tubule cells are limited in their capacity to transfer glucose back into the blood. At a normal blood glucose level of 100mg/100ml, no glucose will be excreted: but if it exceeds 180mg/100ml, glucose appears in the urine. Patients with untreated diabetes mellitus have high blood sugar levels, and consequently have glucose in their urine.

Relationship between concentration of glucose in plasma and the rate of excretion of glucose in urine

(T_m = maximum rate of glucose reabsorption)

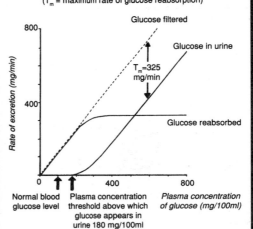

Glucose filtered

Glucose in urine

800

T_m=325 mg/min

Rate of excretion (mg/min)

400

Glucose reabsorbed

0   400   800

Normal blood glucose level

Plasma concentration threshold above which glucose appears in urine 180 mg/100ml

Plasma concentration of glucose (mg/100ml)

Other active reabsorptive systems in the tubules are also only able to deal with limited amounts of material per unit time. This is due to the transport membrane carriers becoming saturated. Various organic solutes and amino acids are reabsorbed by such limited capacity, active transport systems. These systems enable the kidneys to protect against significant loss of the substances, but do not help set their plasma concentrations.

Urea

Urea is passively reabsorbed. It is freely filtered at the glomerulus, and so in the glomerular filtrate, its concentration is the same as in plasma. As the filtrate flows down the tubule, water is reabsorbed and causes the concentration of urea to rise. This creates a concentration gradient from the tubule to the capillary down which the urea diffuses.

Reabsorption of urea

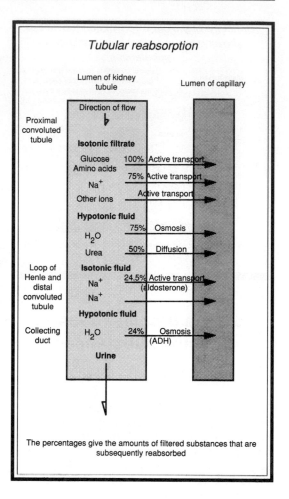

The percentages give the amounts of filtered substances that are subsequently reabsorbed

Tubular secretion

There are three distinct active transport systems used in secretion. Each system shows a low degree of specificity, and so can transport a wide range of substances. Hence they are suitable for eliminating drugs and other foreign chemicals. Some substances such as paraaminohippuric acid (PAH), are so ruthlessly excreted, by both filtration and tubular secretion, that the renal veins are almost completely free of them. Quantitatively the major ions secreted by the tubules are hydrogen, potassium and ammonium ions.

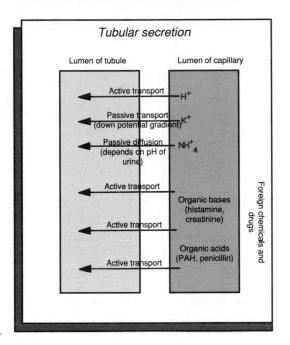

Tubular secretion

Lumen of tubule   Lumen of capillary

Active transport — H^+

Passive transport — K^+
(down potential gradient)

Passive diffusion — NH_4^+
(depends on pH of urine)

Active transport

Organic bases
(histamine, creatinine)

Active transport

Organic acids
(PAH, penicillin)

Active transport

Foreign chemicals and drugs

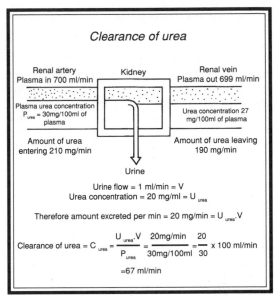

Clearance of urea

Renal artery
Plasma in 700 ml/min

Kidney

Renal vein
Plasma out 699 ml/min

Plasma urea concentration
P_{urea} = 30mg/100ml of plasma

Urea concentration 27
mg/100ml of plasma

Amount of urea
entering 210 mg/min

Amount of urea leaving
190 mg/min

Urine

Urine flow = 1 ml/min = V
Urea concentration = 20 mg/ml = U_{urea}

Therefore amount excreted per min = 20 mg/min = $U_{urea}.V$

$$\text{Clearance of urea} = C_{urea} = \frac{U_{urea}.V}{P_{urea}} = \frac{20mg/min}{30mg/100ml} = \frac{20}{30} \times 100 \text{ ml/min}$$

$$= 67 \text{ ml/min}$$

RENAL CLEARANCE

Kidney function is assessed using the mathematical concept of clearance. If the concentration of substance x in the urine is U_x g/ml and the volume of urine passed is V ml/min, then the amount of substance x leaving the body in the urine per minute will be $U_x.V$ g/min.

If the concentration of substance x in the plasma is P_x g/ml, then the volume of plasma that contains the quantity of substance x excreted in the given time interval will be

$$U_x.V/P_x \text{ ml/min.}$$

This is the clearance of substance x (C_x), and so

$$C_x = U_x.V/P_x \text{ ml/min.}$$

Clearance is measured in the units of a notional volume per unit of time, and is independent of absolute plasma concentration and rate of excretion of a substance. The idea of clearance is best illustrated with a simple example. The calculation in the diagram above right shows that 67ml of plasma contain the amount of urea which passes into the urine per minute.

If there is a substance which is filtered freely by the glomeruli and which passes through the tubules untouched (i.e. neither secreted nor reabsorbed), it would then be possible to estimate the glomerular filtration rate. Inulin is such a substance. When it is injected into the bloodstream, its concentration in the glomerular filtrate is the same as in the plasma, and the amount leaving the kidneys per unit time is the same as the amount originally filtered per unit time.

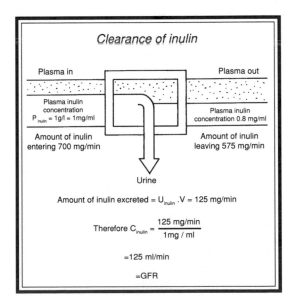

Clearance of inulin

Plasma in

Plasma out

Plasma inulin
concentration
P_{inulin} = 1g/l = 1mg/ml

Plasma inulin
concentration 0.8 mg/ml

Amount of inulin
entering 700 mg/min

Amount of inulin
leaving 575 mg/min

Urine

Amount of inulin excreted = $U_{inulin}.V$ = 125 mg/min

$$\text{Therefore } C_{inulin} = \frac{125 \text{ mg/min}}{1 \text{mg} / \text{ml}}$$

$$= 125 \text{ ml/min}$$

$$= GFR$$

In healthy humans the glomerular filtration rate is remarkably constant over a long period of time. In patients with renal disease, the filtration rate may be reduced, and so it is important to be able to measure it. Creatinine is used for these tests, although its clearance does not exactly equal the filtration rate. However the amount of creatinine added to the urine by secretion is small. Creatinine has the advantage over inulin in that it is a normal constituent of blood and urine, and consequently no intravenous infusion is needed.

Some chemicals when introduced into the bloodstream, are completely removed from the plasma as it passes through the kidneys. The clearance of such a substance is equal to the kidney plasma flow. An example is the organic acid, paraaminohippuric acid, which may be injected into the bloodstream.

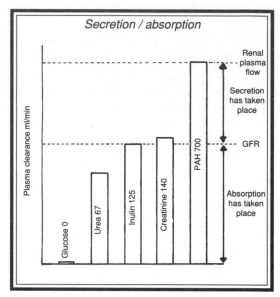

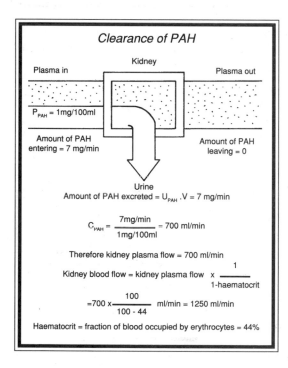

A clearance of a freely filterable substance which is less than that of inulin (125ml/min) indicates that reabsorption of this substance has taken place. A clearance of over 125ml/min means that the substance is being secreted by the tubule cells. This is summarised in the following diagram.

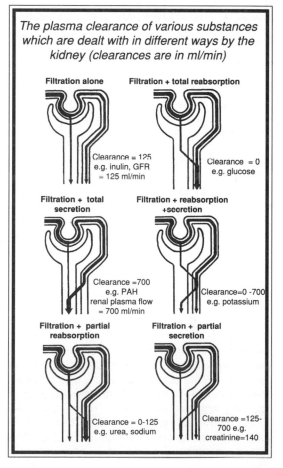

THE ABILITY OF THE KIDNEYS TO PRODUCE CONCENTRATED URINE

In the proximal tubule between 2/3rds and 7/8ths of the water in the glomerular filtrate is reabsorbed. This reabsorption is a passive osmotic event due to the active transport of sodium ions.

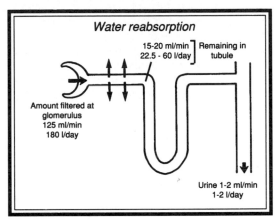

Water reabsorption

15-20 ml/min
22.5 - 60 l/day } Remaining in tubule

Amount filtered at glomerulus
125 ml/min
180 l/day

Urine 1-2 ml/min
1-2 l/day

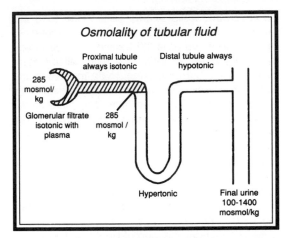

Osmolality of tubular fluid

Proximal tubule always isotonic   Distal tubule always hypotonic

285 mosmol/kg

Glomerular filtrate isotonic with plasma   285 mosmol/kg

Hypertonic

Final urine 100-1400 mosmol/kg

If no further change in tubular fluid volume took place after the urine had passed through the proximal tubule, then the volume would be in excess of 22.5 litres/day (This is the volume voided in diabetes insipidus which occurs when the water reabsorbing mechanisms beyond the proximal tubule fail to function). The maximum volume of urine is, therefore, set by the glomerular filtration rate and the relatively constant reabsorptive activities of the proximal convoluted tubules. However the final urine volume is usually of the order of 1-2 litres/day, and this is determined by events taking place beyond the end of the proximal tubule.

By the time the filtrate has reached the end of the proximal tubule its composition has changed. All the glucose has disappeared due to reabsorption, whilst the concentration of the waste products has risen. However the osmolality is unchanged. It is about 285 mosmol/kg throughout the proximal tubule. The final concentration of the urine can vary considerably, but this again depends on events occurring beyond the proximal tubule. (Definitions of osmotic pressure and osmolality are given in Chapter 3.)

The secret of the kidney's power to concentrate the urine lies in the gradation of tonicity through the medulla, with the highest osmolality at the medullary tip. This osmotic gradient depends on the long hairpin loop shape of the loops of Henle. The hairpin loop allows the system to maintain the osmotic gradients by a counter-current multiplier mechanism.

The action of the loops of Henle relies on the existence in the thick ascending limbs of an active transport system that pumps chloride ions into the interstitium. Sodium ions follow passively.

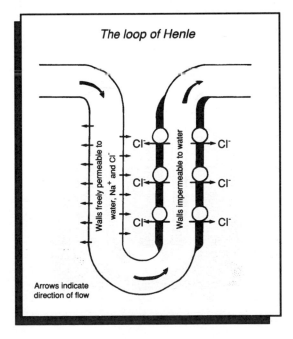

The loop of Henle

Walls freely permeable to water, Na^+ and Cl^-

Walls impermeable to water

Cl^-
Cl^-
Cl^-

Cl^-
Cl^-
Cl^-

Arrows indicate direction of flow

However the walls of this part of the loop are impermeable to water, so as solute is pumped out of the tubule, water is left behind. Hence at any particular level the interstitial fluid will be more concentrated than the urine inside the ascending limb at that level. Furthermore as the urine passes up the tubule, it becomes more and more dilute. Now the descending limb of the loop is freely permeable to water and sodium (and chloride) ions. This means that water will flow out of this part of the loop in order to equalise the osmotic pressure of the lumen of the descending limb with that of the surrounding interstitium. Thus the fluid travelling down the descending limb becomes progressively more concentrated as water is removed at each level. The longer the loop of Henle is, the greater the concentration of the fluid at the base of the loop can be.

In summary the fluid in the descending limbs is continually increasing in osmotic concentration as it flows towards the tip of the medulla, while that in the ascending limbs is continually decreasing in osmolality as it flows back again towards the cortex. Continuous activity of solute pumps in the ascending limbs is required to maintain the concentration gradient.

The blood vessels coursing through the medulla have a hairpin configuration as well. This allows them to act as passive counter-current exchangers. Consequently blood flow through the kidney does not disturb the concentration gradients set up by the loops of Henle.

The ability to concentrate the urine lies in the fact that the collecting ducts plunge down through the medulla amongst the loops of Henle, and pass through the concentration gradient of the medullary interstitium. When water conservation is not needed, the urine remains dilute until it leaves the kidneys, with the same osmolality as in the distal tubules. This occurs because the collecting ducts are effectively "water proof" and so their contents are not affected by the osmotic gradients through which they pass. However, when water conservation is required, antidiuretic hormone (ADH) renders the walls of the collecting ducts permeable to water. The gradient of tonicity in the medullary interstitium results in the removal of water from the collecting ducts. Consequently the urine now becomes more concentrated as it flows down the collecting ducts.

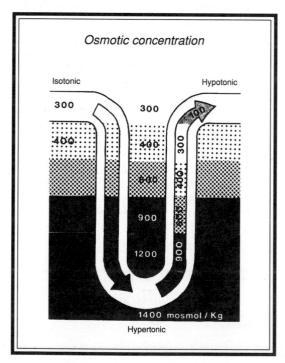

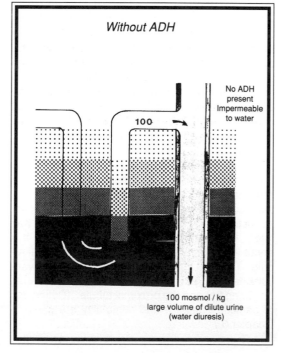

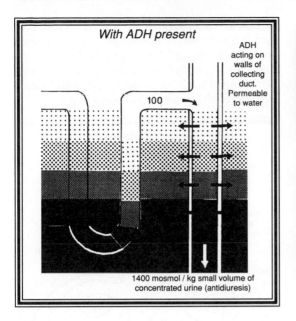

With ADH present

ADH acting on walls of collecting duct. Permeable to water

100

1400 mosmol / kg small volume of concentrated urine (antidiuresis)

Effect of lack of water intake

Small volume of concentrated urine
400 ml / day, 1200 mosmol / kg

SALT AND WATER BALANCE

The kidneys can adjust the volume and concentration of urine over a large range. As a result they are able to maintain the volume and osmolality of the extracellular fluid constant, despite wide variations of fluid intake, climatic conditions, salt intake, etc.

The effect of drinking large volumes

Large volume of dilute urine
25 l / day, 100 mosmol / kg

Control of water excretion

Suppose a person drinks 2 litres of water without altering his total salt content and so, produces a decrease in his body-fluid osmolality. This activates compensatory mechanisms which cause the excess water to be excreted, without altering the renal excretion of salt. Hence the status quo is restored. On a hot day when a person does not have access to fluid, water must be retained by the kidney.

The control of water loss involves the action of ADH on the kidney. There are osmoreceptors located in the brain, which respond to changes in the osmolality of blood plasma. These changes are communicated to hypothalamic neurosecretory cells which release ADH into the bloodstream via the posterior pituitary, in amounts related to the need for conserving water. ADH secretion is also influenced by plasma volume and emotional state (see Chapter 17).

Control of extracellular sodium

The amount of sodium excreted each day must balance the daily intake of sodium.

Routes of sodium chloride intake and loss for a man on an average western diet			
Output	g/day	**Intake**	g/day
Sweat	0.25		
Faeces	0.25	Food	10.5
Urine	10.00		
Total	10.5		10.5

Only sodium loss via the kidneys can be regulated. The rate of excretion of sodium in the urine depends on the rate at which it is filtered at the glomeruli, and on the capacity of the tubules to reabsorb sodium from the filtrate. About 65% of the filtered sodium is actively reabsorbed by the time the filtrate gets to the end of the proximal convoluted tubule. The fraction of the sodium in the filtrate which is reabsorbed remains constant despite changes in glomerular filtration rate. This is called glomerular tubular balance.

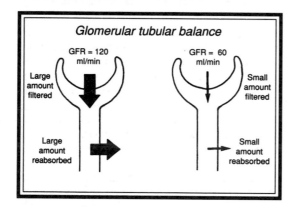

Glomerular tubular balance

GFR = 120 ml/min — Large amount filtered — Large amount reabsorbed

GFR = 60 ml/min — Small amount filtered — Small amount reabsorbed

However the major responsibility for the actual control of sodium excretion lies in the reabsorptive mechanisms in the distal convoluted tubules. Here sodium reabsorption is regulated by the adrenal hormone, aldosterone. Aldosterone is only able to alter the quantity of sodium reabsorbed by 2%. However if this is considered in terms of the amount of sodium filtered each day (1,620g/day), then 2% is 33.4g/day. This is more than three times the amount the average person eats.

The secretion of aldosterone from the cortex of the adrenal gland brings about the retention of sodium by the kidneys. A reduction in aldosterone levels causes sodium excretion. Aldosterone also controls the secretion of potassium ions in the distal convoluted tubules. Levels of aldosterone in the blood are controlled principally by the renin-angiotensin system. Other factors influencing aldosterone secretion are given in Chapter 17.

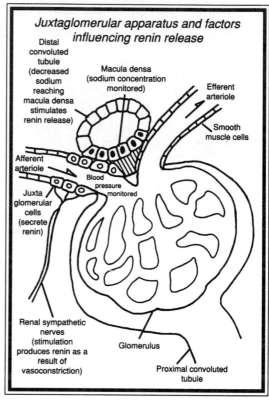

Juxtaglomerular apparatus and factors influencing renin release

Distal convoluted tubule (decreased sodium reaching macula densa stimulates renin release)

Macula densa (sodium concentration monitored)

Efferent arteriole

Smooth muscle cells

Afferent arteriole

Blood pressure monitored

Juxta glomerular cells (secrete renin)

Renal sympathetic nerves (stimulation produces renin as a result of vasoconstriction)

Glomerulus

Proximal convoluted tubule

Renin is secreted into the blood by the juxtaglomerular cells in the kidney. These cells respond to a decrease in blood volume and flow by secreting renin. Renin secretion is also brought about by the activity of the macula densa. The macula densa is a specialised portion of the distal convoluted tubule where it touches the afferent arteriole and glomerulus from its own nephron of origin. The macula densa probably acts by monitoring the early distal tubule fluid composition, in particular its sodium concentration. Renin is also released by ß-adrenergic stimulation in the renal nerves.

The release of renin into the blood stream initiates a series of events that cause the secretion of aldosterone. Renin is a specific protease and it acts on the plasma angiotensinogen converting it to angiotensin I. This is, in turn, converted to angiotensin II which activates aldosterone secretion from the adrenal cortices.

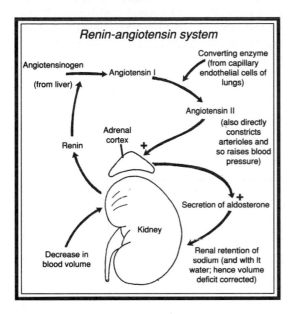

Renin-angiotensin system

Sodium excretion is also controlled by a "third factor". The term "third factor" is used to describe a range of other influences whose precise contributions are at present unclear. They include the redistribution of renal blood flow, physical factors and a natriuretic hormone. The latter increases sodium ion excretion by inhibiting the sodium pumps in the collecting ducts.

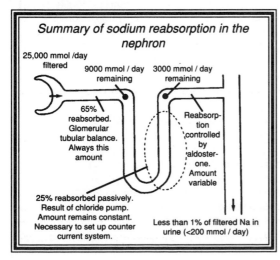

Summary of sodium reabsorption in the nephron

The interrelation of salt and water balance

It should be quite clear that the control of the volume of extracellular fluid and its osmolality is a complicated and integrated process; because of this it is very difficult to consider them in isolation.

The following example illustrates how the systems may cooperate. A healthy individual ingests a large amount of salty food. This excess of sodium chloride produces an increase in osmolality of the extracellular fluid which results in thirst, subsequent drinking and finally an increase in the volume of extracellular fluid (This train of events is often made use of by pub landlords when they provide salty snacks for their customers).

The excretion of salt and water following the drinking of water after a salty meal

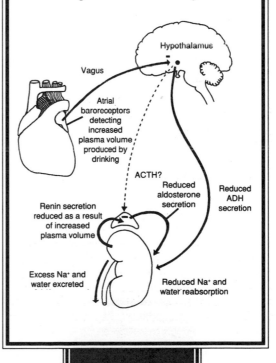

ACID-BASE BALANCE

The pH of body fluids is very carefully regulated.

An acid is defined as a hydrogen ion donor; a base as a hydrogen ion acceptor. When an acid dissociates, it produces hydrogen ions and the conjugate base. Conversely protonation of a base yields its conjugate acid.

$$AH \rightleftharpoons A^- + H^+$$
$$\text{Acid} \rightleftharpoons \text{Conjugate base} + \text{proton}$$

For example an ammonium ion and ammonia are a conjugate acid-base pair.

$$NH_4^+ \rightleftharpoons NH_3 + H^+$$
$$\text{Acid} \rightleftharpoons \text{Conjugate base} + \text{proton}$$

Hydrogen ion concentrations in body fluids are usually expressed in terms of pH.

$$pH = \log_{10}|/[H+] = -\log_{10}[H^+]$$

where $[H^+]$ is hydrogen ion concentration

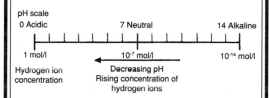

pH scale
0 Acidic   7 Neutral   14 Alkaline

1 mol/l 10^{-7} mol/l 10^{-14} mol/l

Hydrogen ion concentration | Decreasing pH Rising concentration of hydrogen ions

Extracellular fluid has a hydrogen ion concentration of about 40×10^{-9} moles/litre or a pH of about 7.4. Under normal conditions, the pH of extracellular fluid varies very little, and the scale below indicates the range found in healthy individuals. In abnormal situations wider fluctuations may occur.

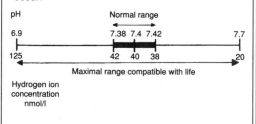

pH   Normal range

6.9   7.38 7.4 7.42   7.7

125   42 40 38   20

Maximal range compatible with life

Hydrogen ion concentration nmol/l

Cells obtain energy by oxidative processes that liberate acidic end products. These acids must be neutralised and subsequently excreted. There are three mechanisms available to do this: buffering, removal of CO_2 from the lungs and excretion of acid by the kidney. They are coordinated to provide precise and continuous control of the pH of extracellular fluid.

1. Buffers

Buffer solutions mop up both excess hydrogen ions and excess hydroxyl ions, and so prevent large changes in pH.

A strong acid such as hydrochloric acid is completely dissociated in solutions at physiological pH.

$$HCl \longrightarrow H^+ + Cl^-$$

Whereas weak acids such as carbonic acid are only partially ionised near pH 7.0.

$$HA \rightleftharpoons H^+ + A^-$$
$$\text{Weak acid}$$

The acid dissociation constant K_A for this reaction is

$$K_A = \frac{[H^+][A^-]}{[HA]} \qquad 1$$

Suppose that some hydrogen ions are added to this solution. In order to continue to satisfy equation 1, some of these hydrogen ions will combine with the salt (A^- in this case) to increase the concentration of the weak acid. In this way the free hydrogen ions are removed from the solution and equilibrium is restored.

The major physiological buffers are:-

• Proteins

Proteins have a large buffering capacity, and are responsible for much of the initial buffering of any pH changes. Haemoglobin, being present in the blood in such high concentrations, plays a major role in this buffering. The buffering properties of proteins are mainly due to amino and carboxyl groups.

$$H^+ + Pr\text{-}COO^- \rightleftharpoons Pr\text{-}COOH$$

$$OH^- + Pr\text{-}NH_3^+ \rightleftharpoons Pr\text{-}NH_2 + H_2O$$

• Phosphates

Phosphates are good buffers but have a limited capacity in the body, since they are only present in rather low concentrations.

• Bicarbonates

The most important buffering system in the body involves bicarbonate ions.

$$H^+ + HCO_3^- \rightleftharpoons H_2CO_3$$

This system is unique as the reaction does not become saturated within the body, and the end products can be dissipated.

$$\underset{\text{Enters general pool}\atop\text{in the body}}{\overset{\text{Carbonic}\atop\text{anhydrase}}{H_2CO_3 \rightleftharpoons}} \overset{\text{Lost via lungs}}{H_2O + CO_2}$$

The enzyme carbonic anhydrase greatly speeds up this reaction.

2. Removal of carbon dioxide via the lungs

When the pH of extracellular fluid falls, more carbon dioxide will be blown off by the lungs. This prevents a build up of carbonic acid in the plasma and enables the reaction described in the diagram to continue to move towards the right. So long as there is sufficient bicarbonate, hydrogen ions will be "mopped up".

Removal of carbon dioxide by the lungs is controlled by the respiratory centre which is in the brainstem. The activity of the respiratory centre is increased when PCO_2 rises and pH of the extracellular fluid falls. In this way, raised PCO_2 and falling pH stimulate ventilation; while decreased PCO_2 or increased pH leads to a lowered ventilation rate.

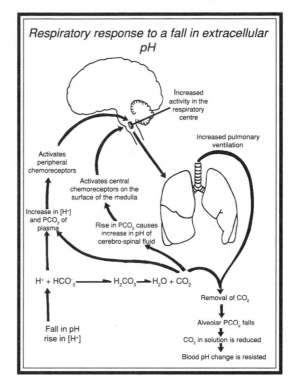

Respiratory response to a fall in extracellular pH

3. Acid excretion by the kidneys

The kidney is capable of excreting urine with a pH outside the normal pH range found in body fluids. Indeed the pH of the urine may vary between 4.5 and 8.2. The kidneys regulate the pH of the body fluid by controlling the supply of bicarbonate.

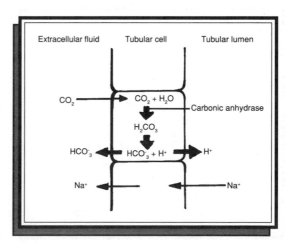

When the pH of the extracellular fluid falls, then CO_2 is produced from the bicarbonate. Carbon dioxide diffuses into the cells of renal tubules and is converted to bicarbonate.

Once hydrogen ions have been secreted into the renal tubules, they are buffered so that a fall in the pH in the lumen is prevented. This buffering in the renal tubules is needed to stop large hydrogen ion concentration gradients building up between the tubular cells and lumen. A large hydrogen ion concentration gradient would effectively reduce further hydrogen ion secretion.

There are three ways in which the tubules buffer the secreted hydrogen ions, and consequently control bicarbonate levels in extracellular fluid:-

• Filtering bicarbonate.

• H^+ ions are buffered by phosphate.

• Formation of ammonia by the tubules.

• Bicarbonate is completely filtered from the plasma at the glomerulus. Normally virtually all the bicarbonate is reabsorbed. The mechanism by which bicarbonate is reabsorbed involves hydrogen ion secretion. When the secreted hydrogen ions meet the filtered bicarbonate in the tubular lumen, they combine to form carbonic acid. The carbonic acid then breaks down to release water and carbon dioxide. These can now diffuse back into the tubular cells to form more bicarbonate and more

hydrogen ions. It must be noted that the bicarbonate ions are unable to cross the membrane between the tubular lumen and the cytoplasm of the tubular epithelial cells. However, in effect, every bicarbonate ion that is trapped by a hydrogen ion in the lumen, produces another in the tubular cells. In this way all the filtered bicarbonate is conserved, and an increase in tubular hydrogen ion concentration is avoided.

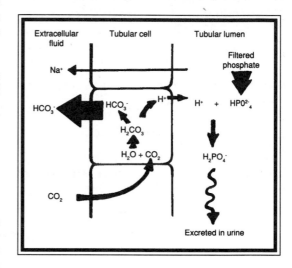

• The secreted hydrogen ions may also be buffered by other molecules filtered at the glomerulus. The most important of these urinary buffers is phosphate.

For every secreted hydrogen ion combining in the

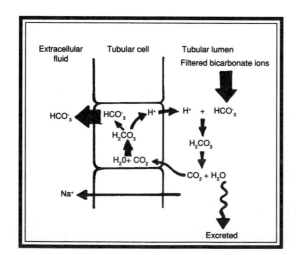

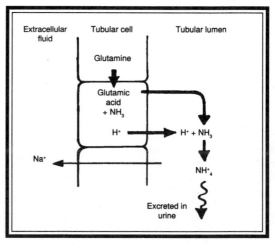

Acidosis and alkalosis

Acidosis occurs when the concentration of hydrogen ions in the plasma rises.

Alkalosis occurs when the concentration of hydrogen ions in the plasma decreases.

Large changes in the PCO_2 following alterations in alveolar ventilation can cause changes in the pH of the plasma (respiratory acidosis and respiratory alkalosis). If the change in plasma pH is due to something other than changes in ventilation then it is referred to as a non-respiratory acidosis or non-respiratory alkalosis.

Non-respiratory acidosis may result from any of the following :-

• Exercise - leading to the formation of lactic acid.

• High protein diet. Sulphur in the protein is oxidised to sulphuric acid and phosphorus is oxidised to phosphoric acid.

• Ingestion of acids e.g. vinegar, acids in soft drinks.

• Untreated diabetes mellitus. Keto acids result from imperfect oxidation of fats.

Non-respiratory alkalosis may result from the following:-

• Ingestion of alkalis such as sodium bicarbonate.

• Vegetarian diet. Salts of organic acids, such as sodium citrate and sodium tartrate, produce basic end products.

• Loss of acid gastric juice by continual vomiting.

In acidosis the kidneys reabsorb all the filtered bicarbonate and also add new bicarbonate to the blood; while in alkalosis bicarbonate is not reabsorbed, but instead it is secreted in the urine.

lumen with a buffer other than bicarbonate, there is a gain of one bicarbonate in the extracellular fluid. This raises the blood bicarbonate concentration causing alkalinisation. At the same time there is the excretion of an equivalent amount of acid ($H_2PO_4^-$) in the urine.

• Buffering of tubular hydrogen ions also occurs by the production of ammonia in tubule cells.

Ammonia diffuses rapidly into the tubular lumen. There it meets secreted hydrogen ions and forms ammonium ions. Unlike the highly diffusible ammonia, the positively charged ammonium ions cannot diffuse back into the cells, and so stay in the lumen.

This system becomes more important as the need to excrete hydrogen ions and regenerate bicarbonate increases. The buffering of hydrogen ions by ammonia has a similar effect to that of phosphate in producing new bicarbonate ions.

NUTRITION

Animals require food to provide energy, and to repair and build their tissues. The energy is needed both for physical activities, such as walking, and for maintaining the internal environment. The minimum daily energy required by the body, when the subject is completely at rest, is called the basal metabolic rate (BMR).

A healthy balanced diet should contain the following classes of substances: carbohydrate, protein, fat, water, vitamins and minerals. The table shows the suggested daily intake of energy and some of these nutrients for a sedentary, young man.

Energy 10-11 MJ	Thiamin 1.0 mg
Protein 60-65 g	Vitamin A 750 µg
Calcium 500 mg	Riboflavin 1.6 mg
Iron 10 mg	Ascorbic acid 30 mg

Dietary requirements vary for different groups, such as growing children, pregnant women or labourers.

CARBOHYDRATE

The main source of energy in the diet is carbohydrates. Their general formula is $C_x(H_2O)_y$. An example of a simple carbohydrate is glucose:-

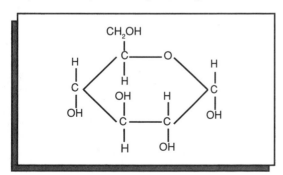

Glucose and other simple sugars such as fructose, are monosaccharides. Monosaccharides can combine to form disaccharides.

glucose + glucose ⟶ maltose (malt sugar)
glucose + fructose ⟶ sucrose (table sugar from sugar cane or beet)
glucose + galactose ⟶ lactose (milk sugar)

When large numbers of sugar molecules polymerise, the product is a polysaccharide. The most important of these is starch which consists of chains of several hundred glucose molecules. Starch occurs as granules in seeds and roots of many plants. Glycogen is the animal equivalent to starch, and is composed of 3,000 to 6,000 glucose units. The main dietary sources of glycogen are liver, meat and shellfish.

Dietary fibre is mainly cellulose and other components of plant cell walls. Cellulose consists of 3,000 or more glucose units. The glucose-glucose linkages in cellulose are different from those in starch or glycogen, and cannot be degraded by gut enzymes. It, therefore, provides bulk to the contents of the gut by virtue of its indigestible nature, and a minimum amount is desirable for normal intestinal function. Dietary fibre also prevents constipation and bowel disorders.

PROTEIN

Amino acids linked by peptide bonds form a polypeptide chain and proteins are composed of one or more of these polypeptide chains. The general formula for an amino acid is:-

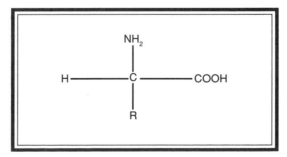

Proteins play key roles in nearly all biological processes, e. g.

• Enzymes catalyze chemical reactions.

• Specific proteins transport and store small molecules.

• Muscle proteins produce coordinated motion.

• Collagen provides mechanical support.

• Antibodies give immune protection.

• Receptor proteins help generate and transmit nerve impulses.

• Growth factors and other proteins control growth and differentiation.

Protein also supplies energy but is not such an important dietary source as carbohydrate or fat.

There are about twenty amino acids found in proteins. The body can synthesise some of these from simple nitrogenous compounds and so they need not be provided in food. These are the non-essential amino acids. The essential amino acids cannot be synthesised in animal tissues and so must be in the diet. Nutritionists assess the biological value of a protein by determining whether it contains all the essential amino acids in the right quantities. Animal foods such as meat, liver, kidneys, milk and eggs are sources of high value protein. Some plant foods such as peas, beans and nuts are also rich in protein, but most plant foods are inadequate protein sources.

FAT (LIPID)

Fats are insoluble in water, but are soluble in organic solvents such as chloroform and methanol. They form the major structural component of membranes, and are also an important source of dietary calories. Indeed fat will yield twice as much energy

as an equal amount of carbohydrate. Excess fat which has not been used to meet immediate energy requirements, is deposited as fat depots in the adipose tissue. These stores are valuable as energy reserves, in maintaining body heat and as padding to protect internal organs.

Most dietary fat occurs in the form of triglycerides which are esters of glycerol and fatty acids. A "fatty" acid is a carboxylic acid with a long hydrocarbon tail. If a fatty acid contains one or more double bonds between carbon atoms, it is unsaturated. Most fatty acids can be synthesised by the body but a few, the essential fatty acids, cannot. All essential fatty acids are polyunsaturated. The main essential fatty acid obtained from plants is linoleic acid. This is converted in the animal body to arachidonic acid. These two fatty acids play a major role in the synthesis of prostaglandins which are vital in the control of many cellular processes. Essential fatty acids are also found in cell membranes and are involved in lipid transport.

Dietary fats are butter, margarine, suet and cooking oils, and fat is also a major constituent of cheese, oily fish, liver, meat and nuts. Essential fatty acids are obtained from many plant oils, but not olive or coconut oil. Provided the supply of fat-soluble vitamins and essential fatty acids are adequate then fat itself is not necessary in the diet.

WATER

Daily intake of water is vital. A person can survive without food for over a month, but without water he will be dead within days.

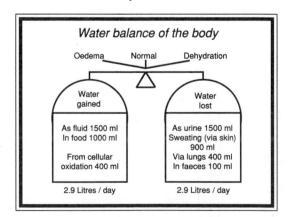

Water intake is governed by thirst. The diagram shows the main factors which are thought to influence thirst through the "drinking centre" in the brain. In turn these factors are influenced by sweating and lack of water intake (and in abnormal situations, vomiting and diarrhoea).

Factors influencing thirst

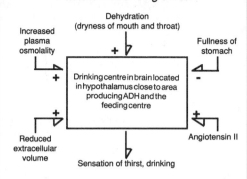

VITAMINS

Vitamins are essential organic molecules which cannot be adequately synthesised by the body. They are only required in small amounts, but their absence from the diet results in well defined diseases.

MINERAL SALTS

More than 20 inorganic minerals are known to be necessary for life. Other minerals are of significance because of their toxic properties e.g. lead, mercury, cadmium and arsenic. There is a constant loss of small amounts of minerals in the urine and other body secretions, and this loss must be replaced. The most important minerals are magnesium, sodium, potassium, chlorine, iron, calcium, phosphorous and iodine.

A summary of the major vitamins

Vitamin	Function(s)	Effects of deficiency	Dietary sources
Fat Soluble			
A Retinol	Visual pigment, maintaining mucous membranes	Night blindness, defects in epithelia	Fish liver oils, milk, butter, cheese, eggs, vegetables, carrots
D	Calcium metabolism	Rickets	Action of UV light on the skin, eggs, liver, oily fish
E	May prevent cell damage	Unlikely in man. Animals : sterility, muscles weakening	Plant oils
K	Blood clotting	Spontaneous or excessive bleeding	Green vegetables, synthesised in bowel by micro-organisms
Water Soluble			
B_1 Thiamin	Carbohydrate breakdown	Beri-beri	Whole grains, yeast, pulses, meat
B_2 Riboflavin	Tissue respiration	Lesions in mouth, tongue, cornea and skin	Yeast, liver, meat, eggs, milk, cheese and vegetables
B_3 Nicotinic acid	Tissue oxidation	Pellagra (nervous disease)	Meat, fish, liver, vegetables, milk, eggs, yeast, can also be synthesised from the amino acid tryptophan
B_5 Pantothenic acid	Constituent of coenzyme A	Unlikely in man	Liver, kidney, egg, yeast
B_6 Pyridoxine	Part of enzyme systems in protein metabolism	Depression and irritability, anaemia	Liver, meat, bran, eggs, vegetables
Folic acid	One carbon transfer reactions	Anaemia	Liver, fish, vegetables, oysters
B_{12} Cyanocobalamin	DNA formation, production of red blood corpuscles	Anaemia	Animal products, liver, meat, fish, milk, eggs
C Ascorbic acid	Metabolism of connective tissues	Scurvy	Fresh fruit and vegetables
Biotin	Some carboxylation reactions	Dermatitis in people eating raw egg white	Liver, kidney, yeast, synthesised in bowel

CONTROL OF FOOD INTAKE

Food intake is controlled by centres in the hypothalamus. If an area of the medial hypothalamus, the satiety centre, is damaged in rats, they eat voraciously and quickly become obese. In contrast the lateral hypothalamus contains a feeding centre, whose destruction leads to rats refusing to eat until they eventually die.

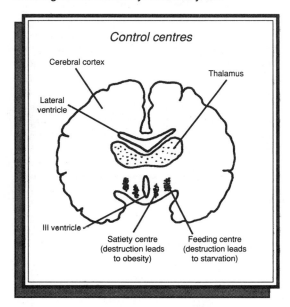

Control centres

Cerebral cortex

Thalamus

Lateral ventricle

III ventricle

Satiety centre (destruction leads to obesity)

Feeding centre (destruction leads to starvation)

A relationship between these centres has been suggested as shown in the flow diagram.

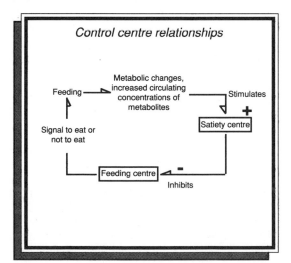

Control centre relationships

Feeding

Metabolic changes, increased circulating concentrations of metabolites

Stimulates

Satiety centre

+

Signal to eat or not to eat

Feeding centre

Inhibits

−

In humans many psychological and social factors also influence food intake. However over the long term a person's weight remains remarkably constant. An eminent physiologist was proud to relate how he could still wear a tailcoat made for him four decades ago despite having subsequently eaten some twenty tons of food. This maintenance of body weight probably involves mechanisms which monitor and regulate body fat. In the short term the factors involved in the control of food intake include:-

• Relaxation of the stomach by eating food

• Solute concentrations of body fluids

• Body temperature regulation

• Blood amino acid levels

• Blood glucose levels

DIGESTIVE SYSTEM

The food we eat is initially in a form which is unable to pass through the walls of the intestine. Consequently it must be broken down into simpler molecules with the aid of hydrolytic enzymes i.e. digested. Absorption through the gut wall can now take place. The products of digestion can then be transported in the bood stream to the cells which can make use of them.

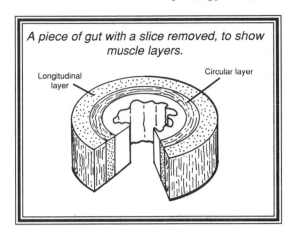

A piece of gut with a slice removed, to show muscle layers.

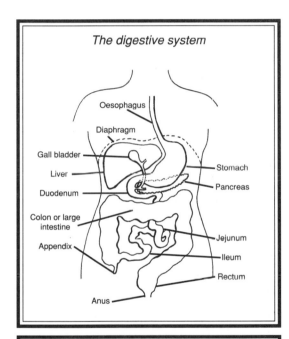

The digestive system

Food is propelled along the digestive tract by alimentary movements consisting of rhythmic muscle contractions and relaxations. The muscular coats of the first third of the oesophagus and the external anal sphincter are striated voluntary muscle. The rest of the alimentary tract is surrounded by two coats of smooth muscle. These smooth muscle layers have a nervous network of their own, which is capable of executing the alimentary movements without any extrinsic innervation. Autonomic nerves may modulate gut activity but they do not command it.

MOUTH

In the mouth food is chewed into small pieces and then swallowed. The presence of food and the act of chewing stimulate secretion of saliva. The rate of secretion increases with larger bites and greater chewing effort. In man 1-1.5 litres of saliva are secreted in a day. Saliva flows from three pairs of salivary glands and numerous small buccal glands, and consists of two types of secretion:-

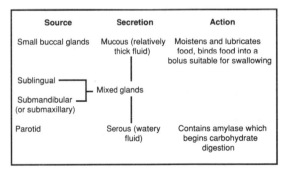

Source	Secretion	Action
Small buccal glands	Mucous (relatively thick fluid)	Moistens and lubricates food, binds food into a bolus suitable for swallowing
Sublingual — ⎫ Submandibular — ⎬ Mixed glands (or submaxillary) ⎭		
Parotid	Serous (watery fluid)	Contains amylase which begins carbohydrate digestion

Saliva keeps the mouth clean and moist in the absence of food and so aids speech. It also dissolves soluble constituents of the food so that flavour is appreciated.

OESOPHAGUS

The oesophagus is a tube connecting the pharynx, at the back of the mouth, to the stomach. On swallowing the food is propelled through the pharynx, and moves down the oesophagus with the aid of gravity and an advancing, "ring-like" contraction of the oesophageal muscles. Such contractions are usually preceded by a wave of relaxation, and are called peristaltic waves. They travel at about 5 cm/sec and take 5-9 sec to travel to the stomach . The movement of food is eased by lubricating mucus. The presence of food in the oesophagus causes the gastro-oesophageal sphincter to relax and allows the food to enter the stomach.

STOMACH

Food is stored in the stomach until it passes at a controlled rate into the duodenum. This overcomes the problem that food is eaten at irregular intervals and in uneven amounts. The stomach is able to accommodate volumes from 1/2 to 5 litres.

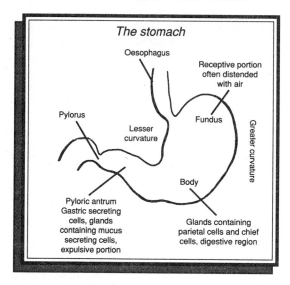

The stomach

The inner surface (gastric mucosa) of the body and fundus of the stomach, is dotted with openings of gastric pits. Glands open into the base of these pits and are composed of three major cell types.

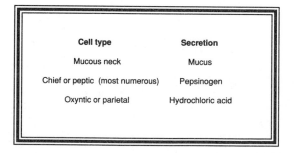

Cell type	Secretion
Mucous neck	Mucus
Chief or peptic (most numerous)	Pepsinogen
Oxyntic or parietal	Hydrochloric acid

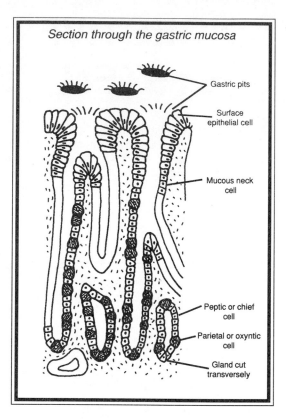

Section through the gastric mucosa

Mixing movements and gastric secretions modify the food in the stomach into chyme, a fairly uniform product as regards pH, osmolality, consistency and temperature.

Gastric movements are waves of muscular contractions originating in pacemaker cells in the longitudinal muscle layer high on the greater curvature. These peristaltic waves spread over the stomach with a speed of 1 cm/sec, and at a frequency of 3/min.

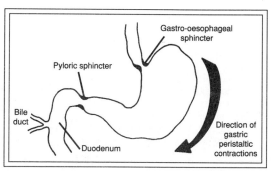

Gastric secretions include hydrochloric acid, pepsinogen, mucus, intrinsic factor and the enzymes, lipase which splits fat and gelatinase which liquifies gelatine.

Hydrochloric acid has the following functions:-

• It breaks down connective tissue and muscle fibres.

• It provides a medium of low pH in which pepsin can act.

• It has a bactericidal action and kills any living cells which have been ingested.

Pepsinogen is the inactive precursor of the enzyme pepsin.

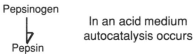

Pepsinogen

In an acid medium autocatalysis occurs

Pepsin

Pepsin breaks down protein molecules by hydrolysing the peptide bonds.

Intrinsic factor is a protein which binds to vitamin B_{12}, and is necessary for its absorption.

The secretion of acid and digestive enzymes is controlled primarily by gastrointestinal hormones (gastrin, PZCCK, secretin, gastric inhibitory peptide). Hormone secreting cells can be found throughout the length of the gastrointestinal tract. These cells are not grouped into distinct glands, but instead they are scattered amongst the mucosal cells lining the gut. They are of many different types, with each type having its own particular distribution and secreting a specific polypeptide hormone.

The gut hormones are mainly secreted in response to changes in the gut contents. The autonomic nervous system can also cause their secretion. These hormones exert their influence either in the classical endocrine fashion or else they have a paracrine action (see diagram). Their main target organs are the gut and pancreas.

The rate at which the chyme passes out through the pyloric sphincter depends on the chemical and physical characteristics of the stomach contents.

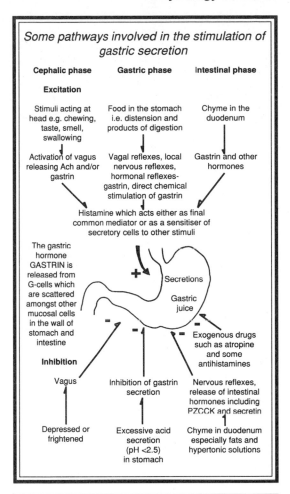

Some pathways involved in the stimulation of gastric secretion

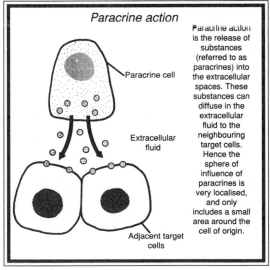

Paracrine action

Paracrine action is the release of substances (referred to as paracrines) into the extracellular spaces. These substances can diffuse in the extracellular fluid to the neighbouring target cells. Hence the sphere of influence of paracrines is very localised, and only includes a small area around the cell of origin.

> The epithelial cells that line the stomach are columnar, mucus-secreting cells, and are joined by tight junctions. This forms the gastric mucosal barrier and enables the stomach to contain acid without injuring itself. The epithelial cells completely replace themselves every three days.

The passage is slowed by solid food, high or low osmolality, fats and acid in the duodenum. Larger volumes of chyme speed up movement.

SMALL INTESTINE

The human small intestine is a tube several metres long. When the chyme first enters the small intestine, its pH and osmotic pressure are adjusted. Then digestion occurs within the intestinal lumen, but it is not always completed there. Many chemical bonds are broken at the mucosal surface of epithelial cells (within the brush border) or even intracellularly. Following digestion, the small molecules are absorbed into the blood stream.

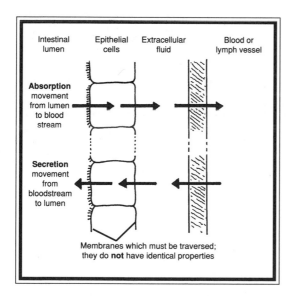

Substances may be transferred in either direction across the epithelial barrier formed by the intestinal walls. The net transfer of a particular substance represents a balance between what is absorbed and what is secreted. Transfer across the intestinal wall may occur by either:-

- simple diffusion

- passive transport

- active transport where energy is required

or

- a combination of these.

Different parts of the small intestine may have different transporting mechanisms, e.g. vitamin B_{12} is absorbed by specific processes in the ileum and not elsewhere.

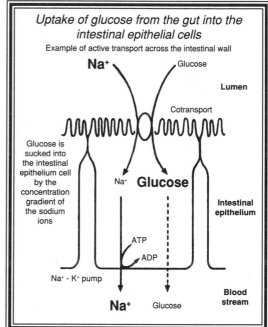

Uptake of glucose from the gut into the intestinal epithelial cells
Example of active transport across the intestinal wall

Movements of the small intestine

The movements of the small intestine serve both to mix the intestinal contents and to move them along towards the colon. The chief movement is segmentation, which is primarily a mixing movement. The circular muscle around the intestine contracts in a series of rings each several centimetres from the other. A few seconds later the contracted portion relaxes and the previously uncontracted portion contracts. These rings of contraction do not move along the intestine and so are distinct from peristal-

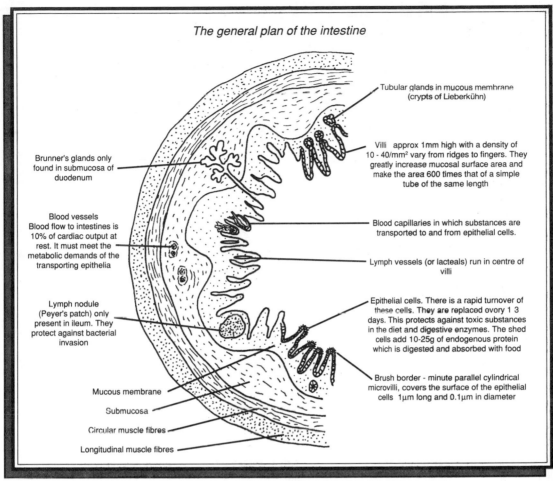

The general plan of the intestine

Tubular glands in mucous membrane (crypts of Lieberkühn)

Villi approx 1mm high with a density of 10 - 40/mm² vary from ridges to fingers. They greatly increase mucosal surface area and make the area 600 times that of a simple tube of the same length

Brunner's glands only found in submucosa of duodenum

Blood vessels
Blood flow to intestines is 10% of cardiac output at rest. It must meet the metabolic demands of the transporting epithelia

Blood capillaries in which substances are transported to and from epithelial cells.

Lymph vessels (or lacteals) run in centre of villi

Epithelial cells. There is a rapid turnover of these cells. They are replaced ovory 1 3 days. This protects against toxic substances in the diet and digestive enzymes. The shed cells add 10-25g of endogenous protein which is digested and absorbed with food

Lymph nodule (Peyer's patch) only present in ileum. They protect against bacterial invasion

Brush border - minute parallel cylindrical microvilli, covers the surface of the epithelial cells 1µm long and 0.1µm in diameter

Mucous membrane
Submucosa
Circular muscle fibres
Longitudinal muscle fibres

sis. Movement of the lumenal contents is mainly accomplished by peristalsis. Peristaltic waves consist of ring-like contractions each preceded by an area of relaxation. They move along the intestine at a slow rate of 1-2 cm/sec, and usually travel only about 10 cm before dying out. Occasionally, in abnormal events such as acute enteritis, peristaltic rushes moving long distances occur.

Secretions of the small intestine

The common bile duct delivers secretions from the pancreas and liver to the duodenum. Other secretions containing mucus come from intestinal glands; Brunner's glands in the duodenum and Lieberkühn's crypts throughout the small and large intestine.

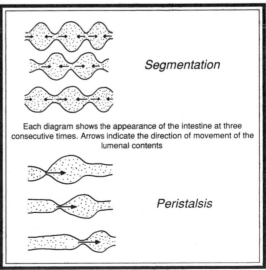

Segmentation

Each diagram shows the appearance of the intestine at three consecutive times. Arrows indicate the direction of movement of the lumenal contents

Peristalsis

Pancreatic secretions

The external secretion of the pancreas contains two components:-

• A large volume secretion with a high concentration of bicarbonate (making it alkaline) is secreted at a rate of 200-800 ml/day in man. A similar fluid (separate from bile acids) is secreted from the liver and is under the same control as that from the pancreas. These alkaline juices neutralise acid entering the duodenum.

• A solution of small volume, containing enzymes, is secreted by the acinar cells of the pancreas.

Some of the enzymes, in particular the proteolytic enzymes, are secreted in inactive forms which are then converted to active forms once in the duodenum. For example trypsinogen is converted into trypsin.

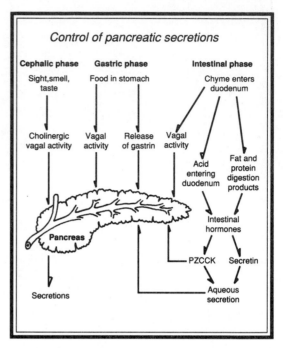

Pancreatic enzyme	Substances hydrolysed
Trypsin Chymotrypsin Carboxypeptidase	Protein
Amylase	Carbohydrate
Lipase Phospholipase	Fat
Elastase	Elastic tissue
Collagenase	Supportive protein of skin and connective tissue
Ribonuclease Deoxyribonuclease	Nucleic acid

Liver secretions

Bile, the neutral, golden-yellow secretion of the liver, is a variable mixture of water, organic and inorganic solutes (also substances metabolised by the liver such as detoxified drugs). The daily output of bile in man is 700-1200 ml. Bile contains the following organic solutes: bile acids, phospholipids (particularly lecithin), cholesterol and bile pigments, chiefly bilirubin.

Bile acids are water-soluble derivatives of cholesterol. Their major roles are:-

• Emulsifying fat by reducing the tension of oil-water interfaces.

• Preventing denaturation of pancreatic lipase as it leaves the surface of emulsified fat droplets.

• Combining with monoglycerides which are the products of fat hydrolysis, to form aggregates called micelles. These micelles dissolve cholesterol, free fatty acids and fat-soluble vitamins.

Seventy five percent of the bile acids synthesised by the liver, are unaltered as they pass along the small intestine. They are reabsorbed unchanged, and then recycled through the liver. The remaining 25% are metabolised by bacteria in the intestine into secondary products. The majority of these are also absorbed, mainly from the ileum and reconverted in the liver into bile acids. However some 500 mg (or 10% of those secreted) of the bile acids are lost each day in the faeces. They are replaced by new bile acids synthesised in the liver, so as to maintain the total amount of bile acids (the bile acid pool) between 2 and 4 g.

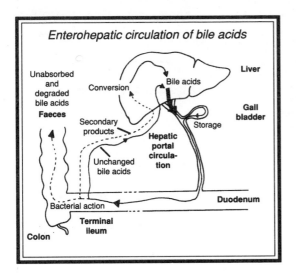

Enterohepatic circulation of bile acids

Control of bile secretion

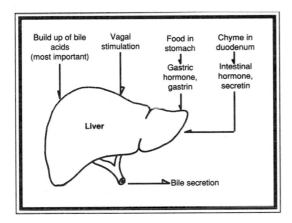

Between meals bile is sequestered in the gall bladder. The gall bladder concentrates the bile, and during the digestion of a meal delivers the bile to the duodenum.

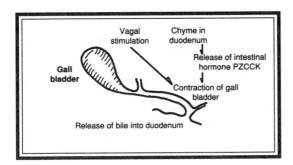

Digestion and absorption of different foods

Substance	Enzyme	Site of action
Carbohydrate (including polysaccharides such as starch and glycogen, and disaccharides such as sucrose and lactose)	Salivary amylase	Mouth to stomach
	Pancreatic amylase	Small intestine
Oligosaccharides (short chains of glucose units) disaccharides	Maltase, lactase sucrase, isomaltase	Brush border of intestinal epithelial cells, greatest activity in jejunum
Monosaccharides (only monosaccharides absorbed)		
Fructose (separate transport system)		Epithelial cells
Glucose, galactose (share sodium dependent active transport system)		Blood capillaries
Triglycerides (emulsification by bile acids and mechanical mixing)	Gastric lipase	Stomach
	Pancreatic lipase	Duodenum
Free fatty acids monoglycerides glycerol form water soluble micelles (4-6 nm dia)		
Micelles are broken down		Brush border in duodenum and jejunum
Absorbed into epithelial cells by diffusion		
Rebuilt into **triglycerides** and form **chylomicrons** (100nm dia) which are secreted into lymphatic system		Within intestinal epithelial cells
		Lymph vessels or lacteals
Phospholipids	Phospholipase from pancreas	Duodenum
Lysolecithin free fatty acids		
Absorbed into epithelial cells		
Glycerol, phosphate, fatty acid	Cytoplasmic enzymes including phosphatases	Cytoplasm of epithelial cells
Protein	Pepsin (acid pH)	Stomach
Proteins, polypeptides		Small intestine
	Trypsin Chymotrypsin	
Peptide fragments		
	Carboxypeptidase and other exopeptidases	Brush border of intestinal epithelial cells
Tripeptides dipeptides amino acids		Lumenal membrane of epithelial cells
absorbed by at least 4 transport systems		
	Peptidases (cytoplasmic enzymes)	Cytoplasm of epithelial cells
Amino acids are usually the products that occur in the blood stream		

Water

In addition to the water in food and drink, a large volume of water is added to the gastrointestinal contents by secretions. The daily volume handled by the gut is 5-10 litres, but less than 100 ml are excreted in the stools. The small intestinal mucosa is highly permeable to water. As a result of this, osmotic equilibrium is quickly established between the contents of the duodenum and the blood.

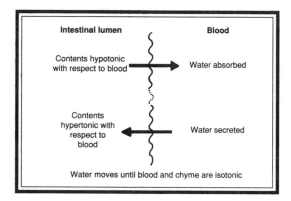

Sodium ions

Sodium ions are absorbed through the walls of the small intestine. The active transport of sodium ions is also coupled to the absorption of many other substances from the intestine, e.g. glucose, some vitamins and amino acids.

Vitamins

Fat soluble vitamins are absorbed like other fats. Most water soluble vitamins are absorbed by active transport mechanisms. A few vitamins are transported by simple diffusion e.g. vitamin B_6 or pyridoxine.

LARGE INTESTINE OR COLON

The ileum is stimulated reflexly as the stomach empties. Its semifluid contents pass into the caecum at the beginning of the large intestine (gastroileal reflex). The colon receives about 1.5 litres each day at an irregular rate. Once in the colon further absorption of salts and water occurs from the residue of digestion.

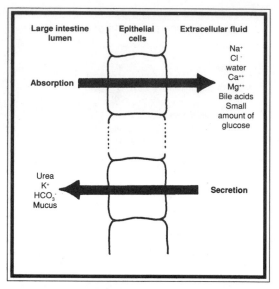

The absorption of sodium ions and the secretion of potassium ions is controlled by the hormone aldosterone.

The colon is inhabited by a vast number of bacteria. These play a vital role in the development and survival of the host.

• They decrease the body's susceptibility to infection. Lymphoid tissue and inflammatory cells are greatly increased by their presence.

• They help degrade endogenous proteins e.g. digestive enzymes, mucoproteins and cellular debris. The products can then be recycled.

• They synthesise some vitamins, e.g. vitamin K, riboflavin and thiamin, and so contribute towards the vitamin requirement of the body.

Slow movements of the colon aid the absorption processes. Regularly spaced, ringlike contractions of the circular muscle of the colon divide it into haustra. The slow contraction is replaced by a slow relaxation, and the neighbouring, hitherto relaxed, ring of circular muscle slowly contracts. The contents of the colon are thus shuttled back and forth. The haustral movements are similar to segmentation in the small intestine, but take several minutes instead of seconds. Occasionally multihaustral segmental propulsion occurs: several haustra contract at once. Peristalsis also occurs moving at a rate of 1-2 cm/min. The net forward progression of all these movements is only 5 cm/hour.

About 36 hours after eating, food becomes the solid or semisolid contents of the large intestine, known as faeces. The main constituents of faeces are:-

• Water (60% - 80% of total weight).

• Nitrogenous material (1-2 g/day), includes bacteria (many millions excreted daily), digestive enzymes, desquamated cells and urea.

• Fatty material, includes fatty acids, neutral fats, phospholipids, sterols and degraded bile acids.

• Fibre (20-60 g/day), includes cellulose and other undigested food residues.

The colour of faeces is due to derivatives of bile pigments. The total weight of faeces excreted each day is between 80 and 500 g, depending mainly on the amount of undigestible fibre eaten.

Motility of the large intestine increases after a meal and then the faeces are pushed into the rectum. The rectal distension arouses the urge to defaecate, the defaecation reflex. This is a local reflex but it is modified by higher centres, so that defaecation occurs at an appropriate time and place. The normal frequency of defaecation ranges from after every meal to once every 3 days.

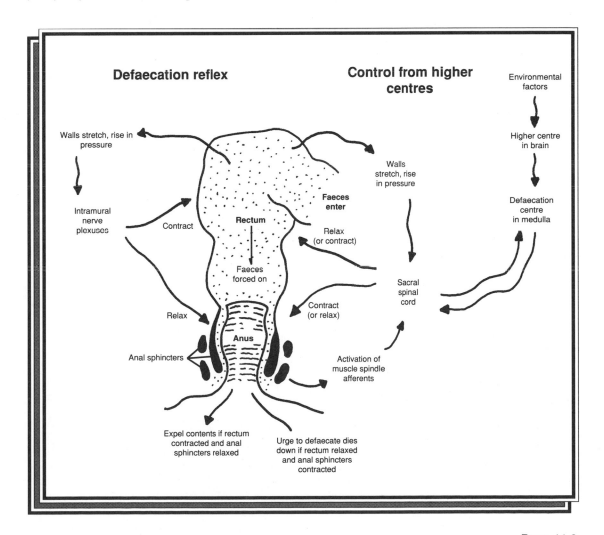

PERIPHERAL NERVE

The nervous system consists of millions of individual nerve cells or neurones, in addition to supporting cells called neuroglia. The neurones are cells which are specialised for conducting nerve signals and have junctions with each other called synapses. This allows nerve impulses to traverse millions of different pathways in the peripheral and central nervous systems.

NEURONES

Neurones are very variable in size and shape. Their typical characteristics are shown in the diagram.

Motor neurones and sensory neurones have the same structure; the former initiate impulses which cause muscles to contract. Sensory neurones conduct impulses from peripheral receptors towards the brain or spinal cord. Some of the axons of both motor amd sensory neurones may be very long indeed compared with the size of the soma. For example, in humans the cells of origin of motor fibres travelling to small muscles in the foot may be more than one metre in length.

Nerve impulses

The properties of the nervous system depend upon the passage of electrical impulses along axons and the transmission of signals across synapses. In this regard there is no difference between impulses in sensory or motor nerves. The difference merely lies in the geographical route of the nerve fibres.

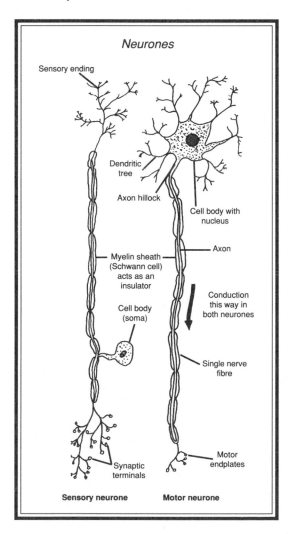

Neurones

Sensory ending

Dendritic tree

Axon hillock

Cell body with nucleus

Axon

Myelin sheath (Schwann cell) acts as an insulator

Cell body (soma)

Conduction this way in both neurones

Single nerve fibre

Synaptic terminals

Motor endplates

Sensory neurone   **Motor neurone**

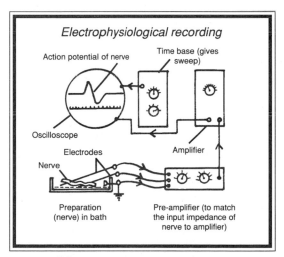

Electrophysiological recording

Action potential of nerve

Time base (gives sweep)

Oscilloscope

Amplifier

Electrodes

Nerve

Preparation (nerve) in bath

Pre-amplifier (to match the input impedance of nerve to amplifier)

Excitable membranes

All cells show a potential difference across their membranes, usually between 10 and 100mV. The reason for this potential difference is an excess of negative charges within the cell and an excess of positive charges in the tissue fluid outside it. These small potentials can be recorded with microelectrodes.

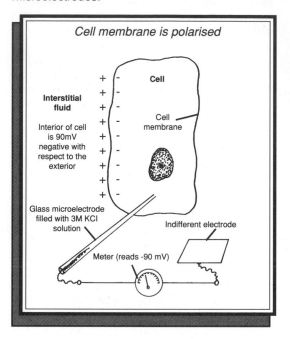

Cell membrane is polarised

Cell

Interstitial fluid

Interior of cell is 90mV negative with respect to the exterior

Cell membrane

Glass microelectrode filled with 3M KCl solution

Indifferent electrode

Meter (reads -90 mV)

RESTING MEMBRANE POTENTIAL

If a microelectrode (an electrode with a tip small enough to be used inside a cell) is inserted through a nerve cell membrane so that its tip lies within the cell, a potential difference of about 80 or 90 mV is registered. The inside of the cell is negative with respect to the outside. This is called the resting membrane potential.

The membrane potential is due to the unequal numbers of ions on either side of the membrane. There are four species of ion that are important as far as membrane potentials are concerned. These are: sodium ions (positively charged); potassium ions (positively charged); chloride ions (negatively charged) and certain organic ions (e.g. protein) which are also negatively charged. Most of the sodium ions remain outside the cell; most of the potassium ions are inside it. Very steep concentration gradients occur across the membrane for all four ions. However, the distribution of these ions is maintained in a steady state.

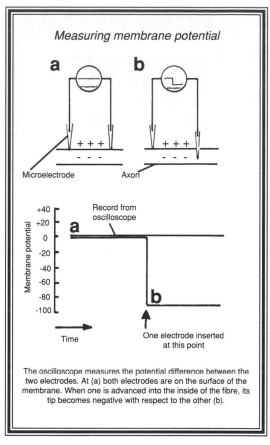

Measuring membrane potential

a   b

Microelectrode   Axon

Membrane potential

+40
+20
0
-20
-40
-60
-80
-100

a

b

Time

One electrode inserted at this point

Record from oscilloscope

The oscilloscope measures the potential difference between the two electrodes. At (a) both electrodes are on the surface of the membrane. When one is advanced into the inside of the fibre, its tip becomes negative with respect to the other (b).

What maintains the asymmetric concentration of ions? How does this unequal concentration produce the resting membrane potential?

The role of potassium ions

The first point to understand is that the cell membrane is permeable to potassium ions. They therefore tend to diffuse out of the cell (nerve or axon) down the steep concentration gradient which is of the order of 150mM (155 mM inside; 5mM outside). When they are inside the cell the potassium ions are balanced by the negatively charged organic ions. However, the organic ions cannot get through the membrane because they are larger. Therefore, potassium ions, in diffusing out of the cell, create an excess of positive charges just outside the membrane, and they leave behind an excess of negative charges within the cell.

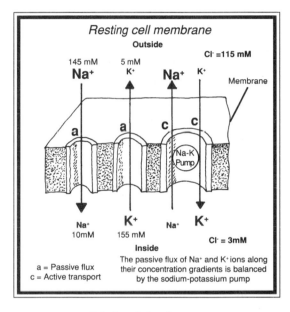

Generation of resting potential

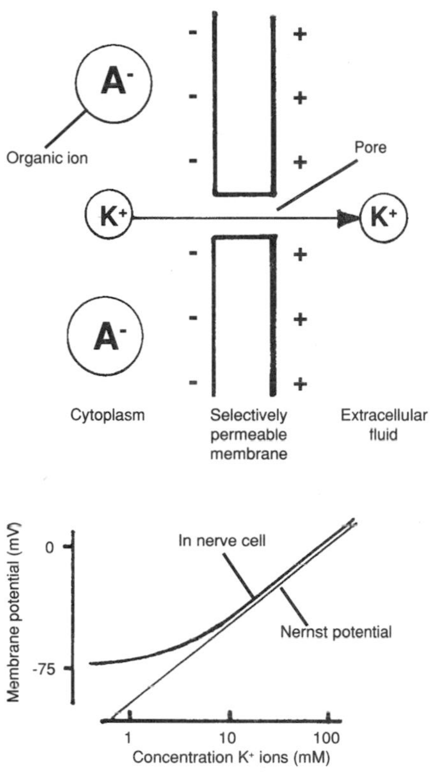

Organic ion

Pore

Cytoplasm

Selectively permeable membrane

Extracellular fluid

Membrane potential (mV)

In nerve cell

Nernst potential

Concentration K+ ions (mM)

Changing K+ concentration alters membrane potential as shown in the graph.

Nernst , in 1888, derived an equation to relate potential and ionic concentration (see Chapter 2).

It can be seen that a state of equilibrium will result when the rate at which potassium ions diffuse out of the axon balances the rate at which the electrical attraction tends to move them back in again. There is also an active metabolically driven pump mechanism in the membrane which moves potassium ions from outside the cell to inside (see Chapter 2). In general, the equilibrium point is reached when the membrane potential is of the order of -70 to -90 mV. The figure depends on the species and on the particular cell type.

Sodium ions

If sodium ions were to travel down their concentration gradient in the same way that potassium ions do, the membrane potential would be removed. However, sodium ions do not penetrate the membrane to any great extent in the resting state. In other words, the membrane is relatively impermeable to them. The small number of sodium ions that do manage to enter the cell are removed from it by the sodium pump (see chapter 2).

Resting cell membrane

Outside

Cl⁻ =115 mM

145 mM Na⁺

5 mM K⁺

Na⁺

K⁺

Membrane

Na-K Pump

Na⁺ 10mM

K⁺ 155 mM

Na⁺

K⁺

Cl⁻ = 3mM

Inside

a = Passive flux
c = Active transport

The passive flux of Na⁺ and K⁺ ions along their concentration gradients is balanced by the sodium-potassium pump

Chloride ions

The membrane is permeable to chloride ions. However, it is at equilibrium when the distribution of ions across the membrane is as indicated in the figure. The reason for this is that the overall electrical attraction caused by the charge difference is just equal to the difference in concentration of chloride ions across the membrane.

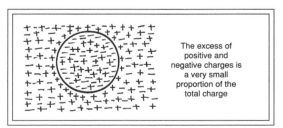

The excess of positive and negative charges is a very small proportion of the total charge

Action potential

When a nerve or muscle fibre is stimulated, an action potential is produced. An action potential can be initiated by anything that results in a depolarisation of the membrane to a certain critical, threshold value. The ionic changes involved in the action potential are:-

- Sodium ions enter the cell.

- Potassium ions leave the cell.

The membrane momentarily becomes very permeable to sodium ions. There are highly developed channels within the membrane, called sodium channels, which open and allow sodium ions to enter the cell. The ions move under the influence of both their concentration gradient and the electrical gradient across the membrane. As the sodium ions enter the cell the membrane potential goes down to zero. The process continues with a reversal of the sign of the membrane potential. The potential in a mammalian axon will rise to +30mV.

Positive feedback

The sodium conductance change is brought about entirely as a result of the initial drop in polarisation of the membrane. The greater the depolarisation, the larger the sodium conductance change and because the sodium ions inside the cell lower the membrane potential still further, then the conductance changes still more. This is an example of positive feedback. In the absence of any cut-off mechanism, this would be a once and for all event. Biologically speaking, positive feedback mechanisms must have a cut-off in order to function more than one time. The cut-off mechanism for the action potential is the so called inactivation of sodium conductance which occurs a set time interval after the sodium channels are opened. It is as if once opened the sodium channel will remain open for a fairly large fraction of a millisecond and then automatically closes, no matter what.

Such positive feedback systems where they occur in biology are examples of biological amplifiers. A very small event, here the initial stimulus to the membrane that depolarises it, produces a large response. There are other examples in biology, such as the blood clotting mechanism (see Chapter 3) and the hormonal control of ovulation (see Chapter 18).

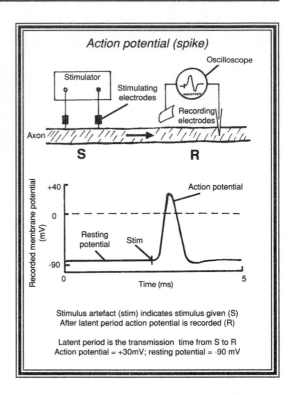

Action potential (spike)

Stimulus artefact (stim) indicates stimulus given (S)
After latent period action potential is recorded (R)

Latent period is the transmission time from S to R
Action potential = +30mV; resting potential = -90 mV

The time dependent inactivation of sodium conductance leads to the closure of the sodium channels, whereupon the membrane becomes more permeable to potassium ions. The cell interior is now charged positively and potassium ions are not held back by electrostatic forces. They then leave the cell through potassium channels driven by both electrical gradients and concentration gradients. As they leave the cell, the positive charge inside is reduced; it falls through zero and back towards re-establishing the resting potential.

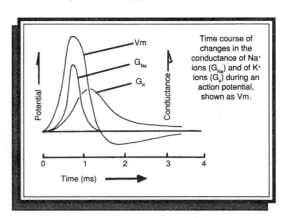

Time course of changes in the conductance of Na$^+$ ions (G_{Na}) and of K$^+$ ions (G_k) during an action potential, shown as Vm.

It should be recognised that only a very small number of ions take part in these exchanges with each nerve impulse, compared with the total number of ions inside and outside the cell. The ions that do take part are soon returned to their rightful places on either side of the membrane by the action of the sodium pump.

The stimulus

The mechanism giving rise to these large effects on the permeability of excitable membranes to sodium and potassium ions is at present unknown. However, many types of stimulus can produce the initial depolarisation. For example, contact with a potassium chloride crystal; raised external concentration of potassium ions; mechanical deformation such as pressure on the nerve with a surgical instrument; and most commonly used in experimental physiology, the application of an electrical stimulus.

Properties of neurones

The conduction velocity in a nerve is the speed at which the impulse travels along it. The larger the diameter of the fibre the faster the nerve impulse will travel. The giant axon of the squid, for example, is part of an escape mechanism and is required for a particularly rapid reflex response. These giant axons are as large as 500 μm across. In mammals, another method has evolved for speeding up conduction because unmyelinated fibres generally reach a conduction velocity limit of 1 or 2 m/sec. This process is known as saltatory (from the latin saltare = to jump). Here the axon is surrounded by a myelin sheath divided into segments which are separated by nodes of Ranvier. The node of Ranvier is bare of myelin and represents a point at which depolarisation of the membrane can take place in response to the action potential existing at a previous node. Conduction therefore proceeds in a series of jumps and, depending upon the distance apart of the nodes which may be as much as 1 mm, the conduction velocity is accordingly increased. The fastest fibres in mammals will conduct at around 120 m/sec. Average human motor fibres conduct at a maximum rate of 52-53 m/sec. Myelinated fibres are used for signals involved in rapidly acting reflexes and muscle control. Unmyelinated fibres conduct impulses to and from the vegetative organs, blood vessels and heart where speed of propagation of impulses is unimportant.

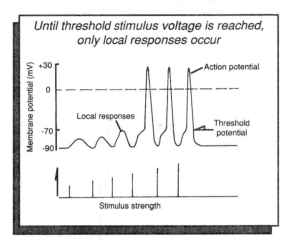

Until threshold stimulus voltage is reached, only local responses occur

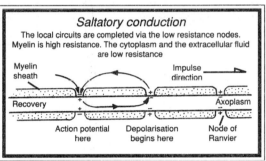

Saltatory conduction
The local circuits are completed via the low resistance nodes. Myelin is high resistance. The cytoplasm and the extracellular fluid are low resistance

The propagation of the action potential

The electrical changes in connection with the opening of the sodium channels also enable the electrical change to move along the nerve or muscle fibre. The cable theory postulates that the cytoplasm inside the axon (axoplasm) has a low electrical resistance, the tissue fluid outside also has a low resistance, whilst the membrane separating them has a high resistance. Therefore, along the length of the axon, a potential difference occurs every time that the membrane is depolarised at a given point. From this point, stretching away on either side, the potential produced by this "little battery", declines with distance (the space constant). At a point distant from the originally active point, the potential across the membrane produced by the completion of the electrical circuit, will be sufficient in turn, to give rise to an action potential. The process will then repeat itself, and in either direction along the axon from the stimulated point, an action potential will proceed at its predetermined velocity, much like a burning fuse of gun powder.

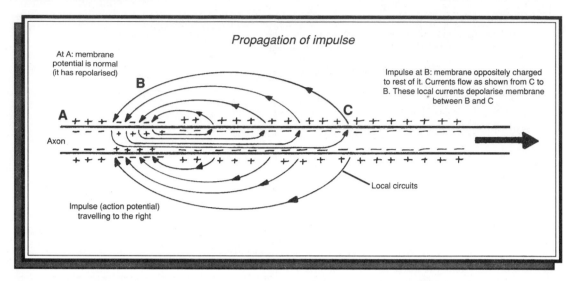

Propagation of impulse

At A: membrane potential is normal (it has repolarised)

Impulse at B: membrane oppositely charged to rest of it. Currents flow as shown from C to B. These local currents depolarise membrane between B and C

Axon

Local circuits

Impulse (action potential) travelling to the right

Refractory period; impulse frequency

When an impulse has passed along a fibre, a time interval takes place in which a second impulse cannot be produced. This is called the refractory period. It is the time required for the membrane potential to return to its resting value. Until this happens, a second action potential cannot be generated. The refractory period, in general, is inversely related to fibre diameter. For the largest fibres it is about 1 ms, which means that these fibres are able to conduct around 1,000 impulses/sec. Smaller fibres usually have refractory periods 10 or 20 times as long as this and, hence, will pass less than 50 to 100 impulses/sec.

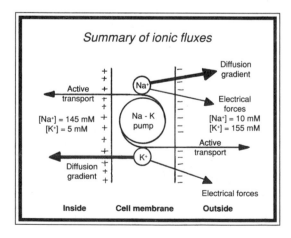

Summary of ionic fluxes

Diffusion gradient

Active transport

Na⁺

Electrical forces

$[Na^+] = 145$ mM
$[K^+] = 5$ mM

Na - K pump

$[Na^+] = 10$ mM
$[K^+] = 155$ mM

Active transport

Diffusion gradient

K⁺

Electrical forces

Inside   **Cell membrane**   **Outside**

Metabolic energy for action potentials

Any axon can produce thousands of impulses without any supply of metabolic energy, before enough sodium and potassium ions accumulate inside and outside the cell respectively, to change the ionic equilibrium so much that action potentials would no longer occur.

When this end point is reached, it is necessary to return the ions to their rightful places before the action potentials can be re-established. This happens as a result of the hydrolysis of ATP which drives the sodium-potassium pump and returns the ions across the membrane. Cells do not require ATP directly in the development of action potentials; once they are initiated, they continue along the fibre without any input of energy. The differential ionic distribution required for the propagation of an action potential is produced by the sodium-potassium pump.

SPINAL REFLEXES

THE NERVOUS SYSTEM

There are two major communication systems in the body. First of all the conduction of information rapidly from one part of the body to another involves the activity of the nervous system. For slower transmission of information the endocrine system is used.

General features

The nervous system consists of central and peripheral divisions. The central nervous system consists of the brain and spinal cord; the peripheral nervous system consists of 12 pairs of cranial nerves which arise from the brain and supply the head and neck region, and 31 pairs of spinal nerves which arise in the spinal cord, pass though openings between the vertebrae and travel through the tissues to innervate the rest of the body.

The cell bodies of all neurones are within the central nervous system or in ganglia. Peripheral nerves consist only of fibres (nerve axons). One spinal

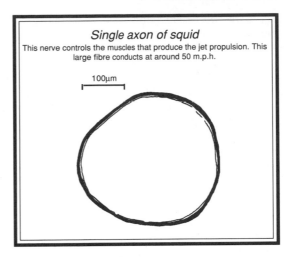

Single axon of squid
This nerve controls the muscles that produce the jet propulsion. This large fibre conducts at around 50 m.p.h.

100μm

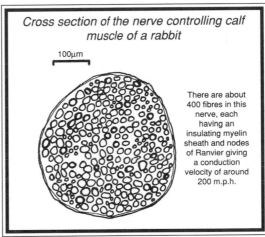

Cross section of the nerve controlling calf muscle of a rabbit

100μm

There are about 400 fibres in this nerve, each having an insulating myelin sheath and nodes of Ranvier giving a conduction velocity of around 200 m.p.h.

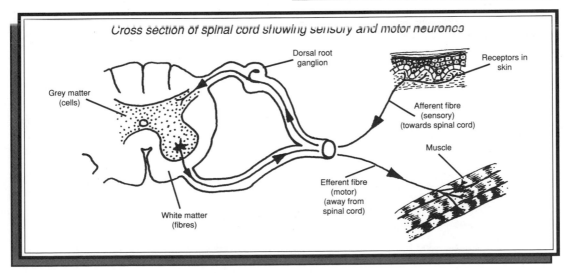

Cross section of spinal cord showing sensory and motor neurones

Dorsal root ganglion

Grey matter (cells)

Receptors in skin

Afferent fibre (sensory) (towards spinal cord)

Muscle

White matter (fibres)

Efferent fibre (motor) (away from spinal cord)

nerve, or one cranial nerve, will have many thousands of individual fibres; some of these carry action potentials from the periphery to the central nervous system, others away from it.

Neurones are connected by synapses that enable nerve impulses to traverse the gap between one nerve cell and another. Nerve fibres in the central nervous system form very complex networks of information channels enabling information to be integrated with other information. The different parts of the central nervous system are connected by ascending tracts, i.e. fibres travelling upwards through the spinal cord from the periphery to the brain, and descending fibres travelling from the brain to muscles and sometimes glands. In the spinal cord and brain, fibre tracts appear to be white and, hence, are termed white matter. Aggregations of cells, often termed nuclei, are much darker in colour and are called grey matter. In the spinal cord the white matter surrounds the grey matter. In the cerebral cortex the grey matter is on the surface and the fibres forming the white matter, are in the core.

SYNAPTIC PHYSIOLOGY

The particular route taken by information within the nervous system is determined by the synaptic organisation. Any one neurone may have many thousands of synaptic contacts from other neurones and, in turn, it will send branches to thousands of other cells.

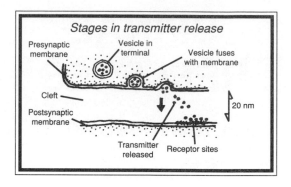

Stages in transmitter release

Synaptic transmission

Axons end in a terminal (synaptic knob). Each terminal contains many little vesicles. A vesicle contains many thousands of molecules of chemical transmitter substance. When a nerve action potential arrives at the terminal, the transmitter is released into the synaptic gap. The transmitter rapidly diffuses across this, and is then bound to receptor sites in the cell membrane of the postsynaptic neurone. When this binding takes place, permeability changes in the postsynaptic membrane. This results in an ionic flux across the membrane,

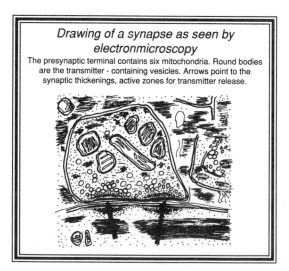

Drawing of a synapse as seen by electronmicroscopy
The presynaptic terminal contains six mitochondria. Round bodies are the transmitter - containing vesicles. Arrows point to the synaptic thickenings, active zones for transmitter release.

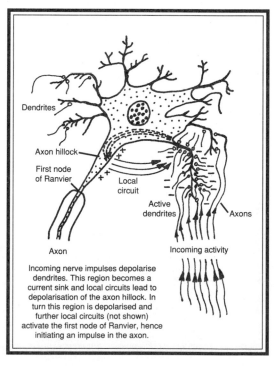

Incoming nerve impulses depolarise dendrites. This region becomes a current sink and local circuits lead to depolarisation of the axon hillock. In turn this region is depolarised and further local circuits (not shown) activate the first node of Ranvier, hence initiating an impulse in the axon.

producing a reduction in the membrane potential of the postsynaptic cell. Such a depolarisation of the membrane is termed an excitory postsynaptic potential or EPSP. As a rule the depolarisation is incomplete, and may range from 5 to 20 mV. If the current-flows within the postsynaptic cell induced by this local, non-propagated depolarisation are sufficiently large, the postsynaptic cell then generates an action potential which travels down its axon away from the soma. The action potential is generated at the axon hillock which is the point at which the axon leaves the soma. It is a process similar to the generation of an action potential in a peripheral nerve.

The transmitter is rapidly removed by destruction by an enzyme or by transfer back into the synaptic terminal. The postsynaptic membrane becomes repolarised and the postsynaptic cell ceases to produce nerve impulses. At least 25 transmitters have been identified and will, when suitably introduced, mimic the effects of nerve impulses. The efficiency of impulse transmission varies with a number of factors:-

• The postsynaptic membrane - some nerve cells have lower thresholds for firing impulses. For example, 5 mV might be a sufficient depolarisation to excite one particular cell, whereas a 50 mV change might be required in others.

• Variability in the amount of transmitter released from single vesicles, and this varies from one synapse to another.

• The synaptic gap which may be of different dimensions; possibly, the wider the gap the less effective transmitters are.

• Various drugs. For example, transmission can be depressed by anaesthetics or enhanced by drugs such as strychnine and caffeine.

Diagram: A synapse

Impulse
Synaptic terminal
Synaptic cleft
Postsynaptic cell
Vesicles
Thickening of membrane

A synapse

Terminal
Vesicle
Cleft
Transmitter
(a) Destroyed by enzyme very rapidly
(b) Recycled back to terminal
Post synaptic membrane (depolarised)

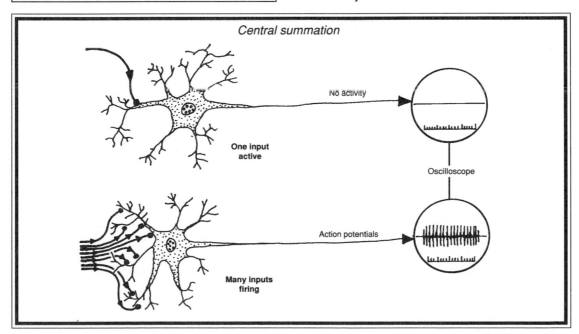

Central summation

One input active

No activity

Oscilloscope

Many inputs firing

Action potentials

Summation

In general, a single impulse reaching a synaptic terminal does not produce a large enough current flow to depolarise the axon hillock to its threshold level. An afferent making synaptic contact with a neurone ends in a number of branches, so that when the fibre is active, several terminals receive the action potentials simultaneously. The combined depolarisation produced is then enough to fire the cell. This phenomenon is known as summation. Summation can be spatial (in space), whereby many afferent nerve terminals are activated at the same time, either because one pathway has a lot of branches ending on the same cell or several pathways are activated at once. Summation can also be temporal (in time), whereby a single terminal, or more usually a group of similar terminals, are repetitively active. In other words, bursts of action potentials reach the ending and the postsynaptic potential attains a level sufficient to exceed the depolarisation threshold. The greater the frequency of incoming action potentials, the larger is the degree of summation.

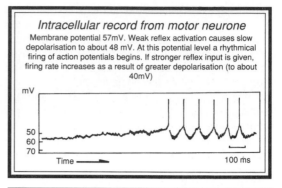

Intracellular record from motor neurone

Membrane potential 57mV. Weak reflex activation causes slow depolarisation to about 48 mV. At this potential level a rhythmical firing of action potentials begins. If stronger reflex input is given, firing rate increases as a result of greater depolarisation (to about 40mV)

mV

50
60
70

Time ⟶   100 ms

One way conduction

It will be evident from what has been said about synapses, that impulses will only travel in one direction across them. In a nerve fibre, or in a muscle fibre, a stimulus applied at any point will produce action potentials propagated away from that point in both directions. However, when a synapse is reached, the impulse can only travel across it in one direction. Normally, transmission is always orthodromic, meaning that it travels in the correct direction. However, antidromic transmission can occur but usually only when physiologists stimulate nerves.

Cell firing rates

The transmission of information in the nervous system depends on what is known as the coding of impulses. Responses are graded in two ways:-

• The number of active nerve fibres in a pathway can be changed. A strong message will activate a large proportion of the available fibres, and a weak message will activate only a small number.

• The frequency of impulse transmission in each fibre can change and strong stimulation results in rapid firing and weak in slow firing.

The frequency of discharge in the postsynaptic neurone then determines the strength of the message. Now, the greater the total amount of transmitter that is released by the synaptic terminals, the greater is the depolarisation produced by local circuit current flow in the axon hillock region. Therefore, the larger the number of synaptic terminals activated (i.e. the greater the summation), the higher will be the frequency of the action potentials generated.

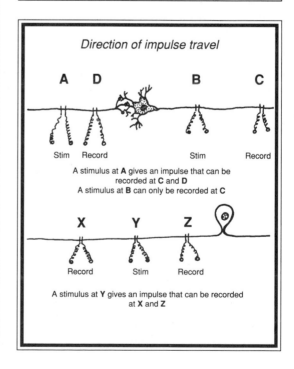

Direction of impulse travel

A   D   B   C

Stim   Record   Stim   Record

A stimulus at **A** gives an impulse that can be recorded at **C** and **D**
A stimulus at **B** can only be recorded at **C**

X   Y   Z

Record   Stim   Record

A stimulus at **Y** gives an impulse that can be recorded at **X** and **Z**

Time course of synaptic transmission

The passage of an impulse across a synapse requires a finite time interval to take place (synaptic delay). This time is much longer than that for transmission along a nerve fibre. In most cases in the mammalian central nervous system, the synaptic delay is about 0.5 ms. This means that in a pathway having large numbers of synapses, the time taken to traverse it will be longer than in a simple one.

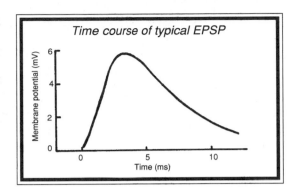

Time course of typical EPSP

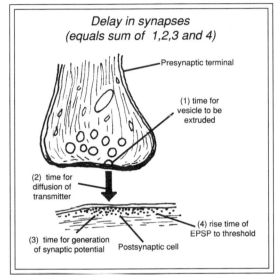

*Delay in synapses
(equals sum of 1,2,3 and 4)*

Inhibition in the central nervous system

The central nervous system probably contains 10^{15} or 10^{16} synapses. Some of these are excitatory, others are inhibitory. By inhibition we mean a situation in which ongoing activity is stopped, or the probability of a response to an excitatory input is reduced. Inhibitory synapses work in roughly the same way as excitatory ones do. Activity in an inhibitory nerve terminal results not in depolarisation of the membrane, but in hyperpolarisation. Hyperpolarisation means that the membrane potential becomes even greater than it was under resting conditions; so, a neurone with a resting potential of -70 mV is hyperpolarised at -80 mV.

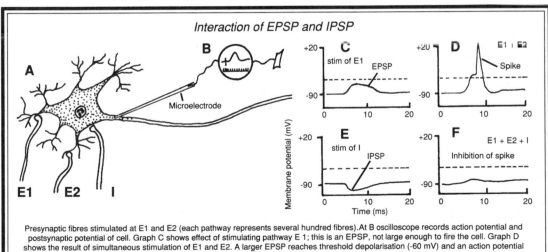

Interaction of EPSP and IPSP

Presynaptic fibres stimulated at E1 and E2 (each pathway represents several hundred fibres). At B oscilloscope records action potential and postsynaptic potential of cell. Graph C shows effect of stimulating pathway E 1; this is an EPSP, not large enough to fire the cell. Graph D shows the result of simultaneous stimulation of E1 and E2. A larger EPSP reaches threshold depolarisation (-60 mV) and an action potential is generated. In E stimulation of the inhibitory pathway, I , gives rise to an IPSP which is a hyperpolarisation of the cell membrane to about -100mV. At F both inhibitory and excitatory pathways are stimulated, but unlike in D the inhibition now prevents cell firing

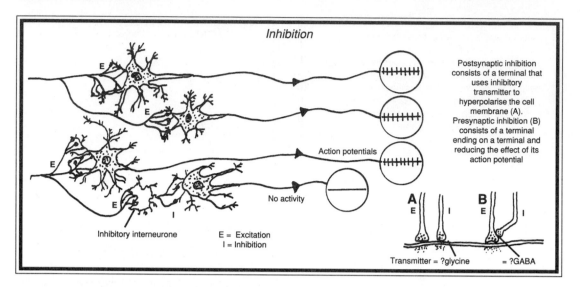

Inhibition

Postsynaptic inhibition consists of a terminal that uses inhibitory transmitter to hyperpolarise the cell membrane (A). Presynaptic inhibition (B) consists of a terminal ending on a terminal and reducing the effect of its action potential

Action potentials

No activity

Inhibitory interneurone

E = Excitation
I = Inhibition

A
E   I

B
E   I

Transmitter = ?glycine   = ?GABA

Firing of the cell is now less likely to happen. In fact, the response of the cell, i.e. its firing at the axon hillock, is the algebraical summation of the activity of all the synapses on that cell. Excitatory ones tend to depolarise the membrane and increase the probability of firing; inhibitory ones have the opposite effect.

Postsynaptic inhibition

Here the action potential arriving at the synaptic terminal releases an inhibitory transmitter. This produces the inhibitory postsynaptic potential (IPSP) which tends to stabilise the membrane, and is accompanied by a slight hyperpolarisation. Any excitation is less likely to be successful in producing an action potential at the axon hillock. Temporal and spatial summation occur in this case as well. The distribution of excitatory and inhibitory synapses differs. Excitatory endings tend to occur mainly on the dendritic tree; inhibitory endings on the base of the dendrites and, to some extent, on the somatic membrane of the cell. The effectiveness of any synapse is related to its distance from the axon hillock. The further away it is, the less effective it is likely to be. Any inhibitory synapses at the extreme ends of dendrites would be ineffective because the excitatory postsynaptic potentials would be generated closer to the cell body than the inhibitory ones.

Presynaptic inhibition

There is also another method of inhibition in the central nervous system. This involves a synaptic influence on the nerve terminal itself as it arrives at the postsynaptic cell. It is possible to have an ending on an ending, and the basic ending will then be affected by the one that is perched on it. This is called presynaptic inhibition because it is the presynaptic membrane that is inhibited as opposed to the postsynaptic one. If the presynaptic inhibitory terminal is activated, it will decrease the amount of transmitter released. Presynaptic inhibition is very potent because its action can completely block the effect of excitatory transmitters. This thereby reduces, and easily prevents, the transmission of impulses from one neurone to another. Strychnine blocks presynaptic inhibition. The effect of strychnine is to produce severe convulsions which are triggered off by the smallest stimulus to any part of the body.

SOMATIC REFLEXES

A reflex is a rapid, automatic, predictable reaction to a stimulus. In general terms, this reaction is directed towards the preservation of the animal. Many reflexes involve the withdrawal from noxious stimulation. The fast removal of the foot from the painful stimulus of treading upon a thumb-tack is an example of the flexor withdrawal reflex.

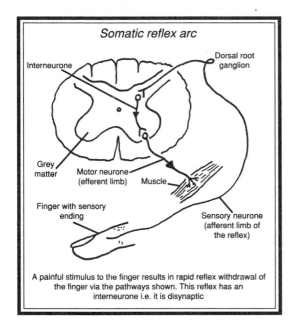

Somatic reflex arc

A painful stimulus to the finger results in rapid reflex withdrawal of the finger via the pathways shown. This reflex has an interneurone i.e. it is disynaptic

trol the response of muscles to stretching; these are involved in the maintenance of posture and in generating movement. Of necessity, they must be rapid in action. These are called monosynaptic reflexes because they do not have an interneurone. The pathway has a single synapse between the afferent and efferent cells.

Looking at the organisation of muscle groups about joints, the simplistic conception of reflexes so far described will not work in practice. The movement of a limb is brought about not by not a single muscle, but by the coordinated activity of several groups of muscles. First of all, there are the prime movers, or agonists, which move the limb around its joint. For this to happen freely, the other muscles acting on the same joint must be inhibited. One can imagine, for example, the biceps muscle which flexes the elbow, requiring the concomitant relaxation of the triceps muscle which extends the elbow. This process is known as reciprocal innervation. It is even more complex than this, because the majority of movements require a fixed base from which to operate. The flexion of the elbow would be of relatively little use in lifting something , if the wrist and shoulder joints were not fixed. Hence, a whole complicated pattern of activity occurs for simple flexion of the elbow.

The impulses involved in reflex activity pass along a reflex arc. Normally, reflex arcs consist of a receptor, an afferent nerve pathway going to the central nervous system, a nerve centre, and an effector pathway.

In the flexor withdrawal reflex, the impulses travel through a comparatively simple pathway. As a rule, there are not more than two synapses in the spinal cord involved. The interposed neurone is called an interneurone. There are other reflexes which con-

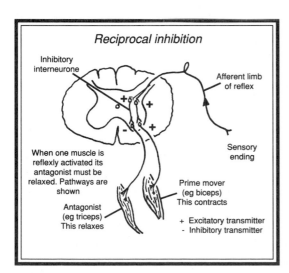

Reciprocal inhibition

When one muscle is reflexly activated its antagonist must be relaxed. Pathways are shown

+ Excitatory transmitter
- Inhibitory transmitter

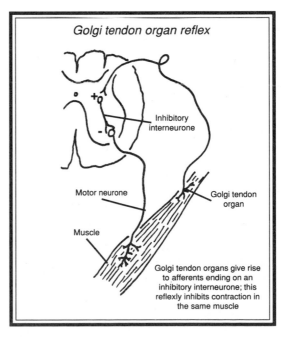

Golgi tendon organ reflex

Golgi tendon organs give rise to afferents ending on an inhibitory interneurone; this reflexly inhibits contraction in the same muscle

CENTRAL NERVOUS SYSTEM

ANATOMY OF THE BRAIN AND SPINAL CORD

The top end of the brain stem is enlarged to form a pair of swellings, the cerebral hemispheres. These have a central cavity filled with cerebrospinal fluid and overhang the brain stem. Below is the cerebellum (part of the hind brain). They have their grey matter or cortex, on the outside, and the white matter within. The cerebral cortex and cerebellar cortex, being convoluted, have very large surface areas. Although the cortex is only about 2 mm thick there are 10^{11} neurones within it. The cerebral cortex is the dominant part of the central nervous system and is the region responsible for thought, memory, language and the various other mental processes which are so highly developed in the human. Nerve fibres, organised into tracts or pathways, connect one part of the central nervous system with another and the two sides of the brain with each other. Cranial nerves arise from the underside of the brain stem. The first two, olfactory and optic, deal with smell and vision; they come from the forebrain. The occulomotor nerves control eye movement and arise in the mid-brain. The other cranial nerves arise from the hind brain and innervate the head and neck. They are involved in eye movements, movements of the face and in speaking. The vagus, or tenth cranial nerve, is the parasympathetic innervation for smooth muscle and glands of the viscera.

Cerebrospinal fluid

The brain and spinal cord, being contained within unyielding bony cavities of the skull and vertebral column, need some kind of protection. This is achieved by the presence of a bathing fluid termed cerebrospinal fluid, which acts as a shock absorber and as a medium for transferring substances between the circulation and nerve cells. This fluid is contained in the ventricles of the brain, the canal of the spinal cord and between the two outer membranes of the brain. Composition of the cerebrospinal fluid is similar to plasma, but contains very little protein and less glucose, potassium and calcium.

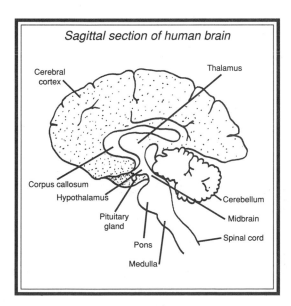

Sagittal section of human brain

Cerebral cortex
Thalamus
Corpus callosum
Hypothalamus
Pituitary gland
Pons
Medulla
Cerebellum
Midbrain
Spinal cord

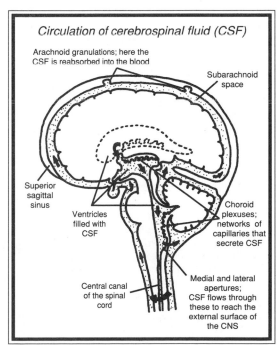

Circulation of cerebrospinal fluid (CSF)

Arachnoid granulations; here the CSF is reabsorbed into the blood

Subarachnoid space

Superior sagittal sinus

Ventricles filled with CSF

Choroid plexuses; networks of capillaries that secrete CSF

Central canal of the spinal cord

Medial and lateral apertures; CSF flows through these to reach the external surface of the CNS

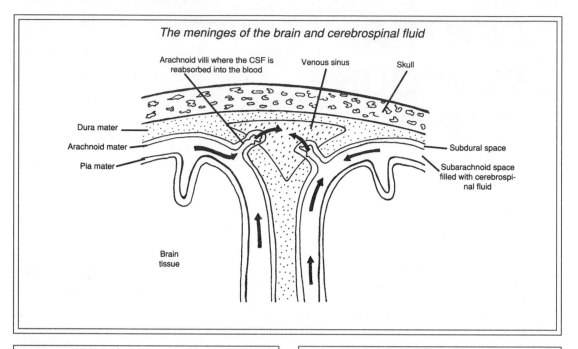

The meninges of the brain and cerebrospinal fluid

Arachnoid villi where the CSF is reabsorbed into the blood

Venous sinus

Skull

Dura mater

Arachnoid mater

Pia mater

Subdural space

Subarachnoid space filled with cerebrospinal fluid

Brain tissue

Blood brain barrier

Since nerve cells are sensitive to very small changes in composition and ionic concentrations of their bathing fluid, it is particularly important that they should be maintained isolated from the fluids circulating in the rest of the body. For this reason there is what is known as the blood brain barrier separating the constituents of blood and plasma from the cerebrospinal fluid. Many substances will readily pass out of capillaries into tissue fluid. In the brain this is not so, because the blood brain barrier blocks them. The capillary membranes in the brain are less permeable than in the rest of the body and the filtration of many substances is prevented e.g. proteins.

The neurones in the brain are supported by specialised cells called neuroglia. They constitute about half of the cellular material within the central nervous system. They have the same function as fibrous tissue in the rest of the body; they are the structural matrix which support nerve cells. They also have other functions, for example some synthesise myelin sheaths, others are responsible for the blood brain barrier, and yet others are phagocytic.

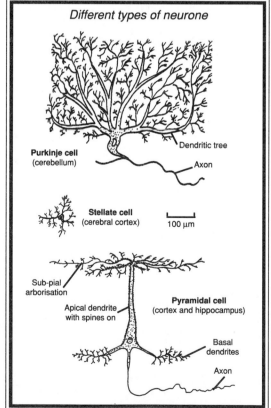

Different types of neurone

Purkinje cell (cerebellum)

Dendritic tree

Axon

Stellate cell (cerebral cortex)

100 μm

Sub-pial arborisation

Apical dendrite with spines on

Pyramidal cell (cortex and hippocampus)

Basal dendrites

Axon

Sensory pathways

All of our waking lifetime, and also during sleep and anaesthesia, sensory information is being transmitted into the central nervous system in vast amounts along pathways which extend from receptors in the skin, muscles, joints, viscera and special sense endings. This barrage of sensory input is sometimes made use of to give a motor response or it may be stored, but the bulk of it is cut off before it ever reaches the brain and consciousness.

This sensory information may or may not give rise to any conscious sensation. This depends to some extent on its origin; if it is from viscera or stretch receptors within muscle, it is unnecessary for it to reach consciousness. On the other hand, visual and auditory information must be closely monitored for purposes of survival.

The sensory pathways have many branches within the spinal cord. Many neurones here are reached by incoming impulses and transmit them to higher levels. Some of the information takes part in reflex action, and some is transmitted to the brain. Other branches sometimes travel to the cerebellum, which is a part of the brain involved in the control of movement. Further branches go to a region known as the reticular formation which controls the state of alertness. Yet other pathways subserve the sensation of pain.

Ascending pathways to the brain

Sensory input travels to the thalamus and the cerebral cortex according to its type. In the spinal cord the fibres subserving the same modality of sensation are gathered into tracts. For instance, sensory fibres from touch receptors enter the spinal cord and then pass to the brain in tracts within the white matter.

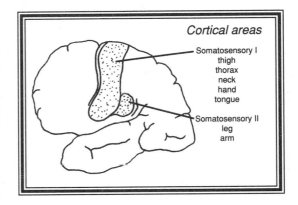

Cortical areas

Somatosensory I
thigh
thorax
neck
hand
tongue

Somatosensory II
leg
arm

All the fibres in these pathways travel to the thalamus before they are relayed to the cerebral cortex. There are at least three orders of neurone in the pathway from the periphery to the cerebral cortex. The first one is in the afferent nerve, the next is in the spinal cord and the third in the thalamus. There is one exception to this, the olfactory pathway.

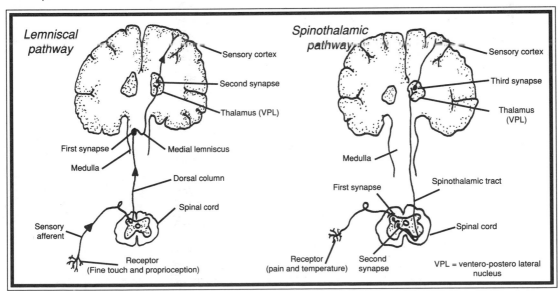

Lemniscal pathway

Sensory cortex
Second synapse
Thalamus (VPL)
First synapse
Medial lemniscus
Medulla
Dorsal column
Spinal cord
Sensory afferent
Receptor
(Fine touch and proprioception)

Spinothalamic pathway

Sensory cortex
Third synapse
Thalamus (VPL)
Medulla
First synapse
Spinothalamic tract
Spinal cord
Receptor
(pain and temperature)
Second synapse
VPL = ventero-postero lateral nucleus

Most sensory pathways cross the mid-line of the spinal cord and travel up on the side opposite to that of the primary receptor. Impulses arising on the right side of the body are transmitted to the left side of the brain i.e. left thalamus and left cortex.

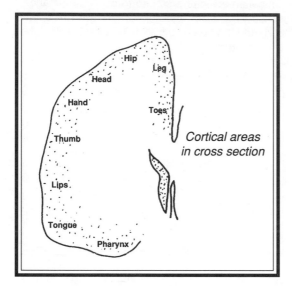

Cortical areas in cross section

The thalamus is not just a relay station; information is changed and integrated with other information and very often is blocked all together. The pattern of action potential output from the thalamus to the cortex is often quite different from the nature of the input. Modification occurs at the other synapses in the pathway as well. The modifications are usually inhibitory and act as filters. If all receptors that are active in the body at a given time produced information which eventually reached the cortex, the cortical neurones would be completely overloaded. If the thalamus is removed, either by injury or by experimental intervention, there will be loss of all sensation. However, the visual, olfactory and auditory inputs would still be intact.

Sensory cortical areas

There are four main lobes in each cerebral hemisphere. These are the frontal lobes, parietal lobes, occipital lobes and temporal lobes. Millions of nerve fibres connect these areas with each other. There is a bridge between the two cerebral hemispheres, left and right, called the corpus callosum and fibres in this coordinate the activity of the right

and left sides of the body. However, complete destruction of these fibres does not give rise to as great a disability as might be predicted from the numbers of fibres involved.

The fibres in the sensory tracts travel in two major pathways. The first is the dorsal column pathway which ascends on the same side of the spinal cord to a relay station in the nucleus cuneatus or gracilis and thence via the thalamus to the cerebral cortex. This pathway is involved in fine touch and proprioception. The second pathway is the spinothalamic, in which the sensory fibre enters the spinal cord, and has a synapse at the same level as it enters. The tract then crosses the mid-line and travels up on the opposite side, the next synapse being in the thalamus. Fibres then travel from the thalamus to the sensory cortex. This pathway is concerned with information from temperature, pain, deep pressure and touch receptors in the skin.

A major sensory area of the cerebral cortex is the primary visual area in the occipital lobe. Here, information from the retina is processed. Then

Drawing to show the projection of each half of the visual field to the cortex. The left half of the visual field projects onto the nasal half of the left eye and the temporal half of the right eye. Since the axons from the nasal half of the retina cross the midline, the information from the left half of the visual field for each eye is combined in the right hemisphere.

Eyeball

Optic nerves

Ganglion cells in retina

Optic tract

Optic chiasma

Lateral geniculate body

Geniculo-cortical tract

Superior quadrigeminal brachium

Visual cortex

there is the primary auditory area. Information from each ear travels to the primary auditory receiving areas in the upper part of the temporal lobes of both hemispheres. These areas analyse features of auditory information such as pitch and rhythm. The third main sensory area is the primary somatic sensory area. It lies in the parietal lobe immediately behind the central fissure which separates the frontal from parietal lobe. Here impulses from somatic receptors (skin, muscles, tendons, joints) are received from the opposite side of the body.

Localisation within the brain

The brain can not only sense the stimulation of, for example, touch receptors in the skin, it can also locate accurately that part of the body being stimulated. The way in which this localisation is carried out is by representing the areas of the body in a geographical map. The sensory pathway from a receptor to its termination in the cerebral cortex remains the same throughout life. Impulses arriving at a particular part of the somatosensory cortex indicate that receptors in the corresponding region of the body have been activated.

In experimental animals, and indeed in humans at operation under local anaesthesia, these somatotopic maps have been investigated. An electrode stimulating a particular part of the sensory cortex, gives rise to sensations which appear to the subject to be at the corresponding part of the peripheral anatomy. For example, it is possible to stimulate the thumb area and a peculiar sensation like pins and needles in the thumb is generated. It has been found that the body is represented on the surface of the cortex upside down. Such topically organised maps are known as homunculi.

Damage to a localised sensory region of the cerebral cortex causes sensory loss in a part of the body (small lesions of the occipital cortex cause a patch of blindness). In man, lesions of the somatic area for the hand result in loss of all delicate discriminatory skin sensations which enable objects to be recognised. Larger lesions render the hand useless. The hand is not paralysed but the patient does nothing with it, so it appears to be.

Sensory association areas

Adjacent to the primary sensory area is the association area to which the sensory area projects information. The association areas are thought to involve the interpretation of sensation, for example, information about the shape, weight, size and nature of objects, the positions of limbs are appreciated and the relationship between limb and object position is registered. Rather little is known about the function of this area.

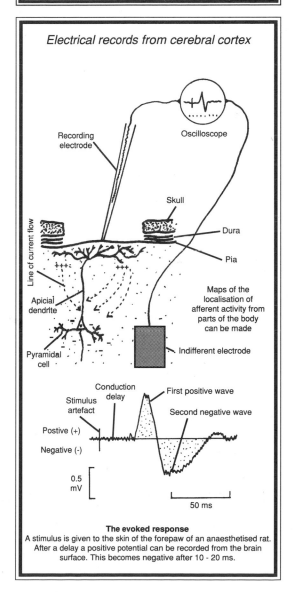

Electrical records from cerebral cortex

Oscilloscope

Recording electrode

Skull

Dura

Pia

Line of current flow

+++

+++

Apicial dendrite

Pyramidal cell

Maps of the localisation of afferent activity from parts of the body can be made

Indifferent electrode

Conduction delay

Stimulus artefact

First positive wave

Second negative wave

Postive (+)

Negative (-)

0.5 mV

50 ms

The evoked response
A stimulus is given to the skin of the forepaw of an anaesthetised rat. After a delay a positive potential can be recorded from the brain surface. This becomes negative after 10 - 20 ms.

HIGHER MOTOR CONTROL

MOTOR PATHWAYS

Movements result from contraction of muscles, the ends of which are attached to the bony skeleton. Impulses controlling muscles travel from the motor cortex to the spinal cord and then in peripheral nerves to the muscle. The descending tracts are of two kinds; both terminate in the spinal cord on motor neurones (alpha efferents) or on the small motor fibres (gamma efferents going to muscle spindles).

Corticospinal tract

The corticospinal tracts run in the medullary pyramids. Neurones extend from the motor cortex to the anterior horn of the spinal cord. Branches go to most areas of the brain stem. The pyramidal tracts cross to the opposite side of the spinal cord. Cutting the corticospinal tracts gives rise to loss of fine skilled movement; gross actions of muscle and maintenance of posture are unaffected.

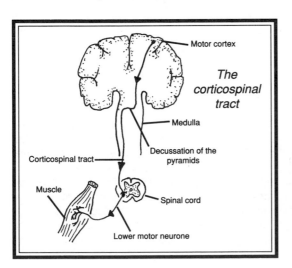

The corticospinal tract

Motor cortex
Medulla
Decussation of the pyramids
Corticospinal tract
Muscle
Spinal cord
Lower motor neurone

Extra-pyramidal tracts

Multineuronal pathways, some starting in the cerebral cortex, others at lower motor centres, travel via several synapses through the brain stem and down the opposite side of the spinal cord. Impulses in these pathways are mostly concerned with posture, background activity for movement, and control of rhythmic movements such as running.

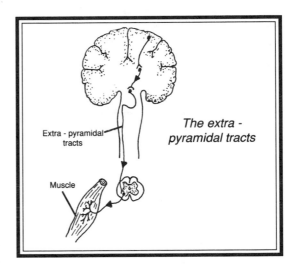

The extra-pyramidal tracts

Extra - pyramidal tracts
Muscle

Under normal circumstances movement involves activity in both these pathways.

MOTOR CONTROL

To move part of the body, various muscles must be activated at the same time (synergistically) and some need to act antagonistically. The biceps and triceps in the upper arm, for example, have antagonistic actions.

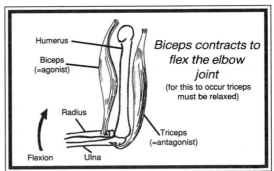

Biceps contracts to flex the elbow joint
(for this to occur triceps must be relaxed)

Humerus
Biceps (=agonist)
Radius
Flexion
Ulna
Triceps (=antagonist)

Motor unit

A motor unit is a motor neurone and the muscle fibres that are innervated by it. The contraction of any muscle is determined by activity in the motor nerve supplying it. In most muscles the motor unit consists of one nerve fibre which branches and supplies about 100 muscle fibres. Muscles involved in skilled movement have a smaller number of muscle fibres supplied by each motor nerve; relatively coarse muscles, e.g. biceps, have motor units consisting of many muscle fibres supplied by one nerve fibre. Any movement is carried out by the activity of one or more motor neurones coordinated within and between different muscles.

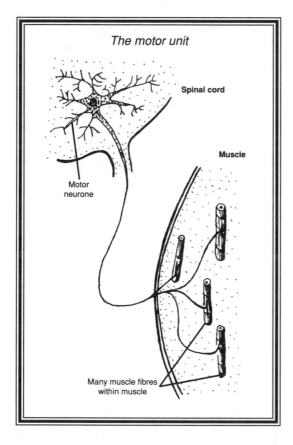

The motor unit

Spinal cord

Muscle

Motor neurone

Many muscle fibres within muscle

The activity originating at the spinal cord level gives rise to simple movements such as reflexes. At higher levels in the central nervous system, more complex movements are generated; the main motor areas are in the cortex and basal ganglia.

Stretch reflex

The basis of all muscle movements is the stretch reflex. Upon the stretch reflex is superimposed activation which originates from other reflexes and higher centres. The stretch reflex consists of an afferent ending (the muscle spindle), an afferent fibre, a reflex centre (in the spinal cord), an efferent fibre and a terminal of the efferent fibre on the appropriate muscle. Muscle spindles lie in parallel with the main muscle and are sensory receptors which respond to stretch. Thus, if a muscle is stretched, by pulling on its tendon or by a movement of the limb, the spindles are stretched. They respond by initiating impulses at an increased rate in their afferents. These impulses travel via the spinal cord back to the same muscle, stimulating it to contract. Thus the basic function of the stretch reflex is to maintain the length of a muscle more or less constant in the face of altering external stretching.

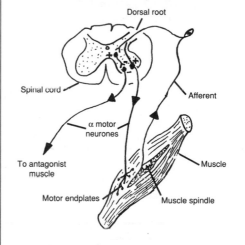

Dorsal root

Spinal cord

Afferent

α motor neurones

To antagonist muscle

Muscle

Motor endplates

Muscle spindle

If the muscle is stretched, the spindle fires and reflex contraction of the muscle occurs. Status quo is maintained. The reflex pathway is monosynaptic. There are no interneurones. This ensures that the stretch reflex is quick.

Stretch reflexes are in action continuously. They are responsible for the resting tone (tonus) in muscles. They also oppose the effect of gravity. Therefore stretch reflexes are of importance in the maintenance of an upright posture. If a forward deviation occurs from the standing posture, muscles in the back and down the back of the leg are stretched. The stretch reflex ensures that a reflex response occurs in these muscles which restores the body to the upright position.

Depending upon the required posture such as standing upright, bending down, kneeling etc., muscles can be maintained at a large number of varying lengths. The stretch reflex is adjustable to work under these different circumstances.

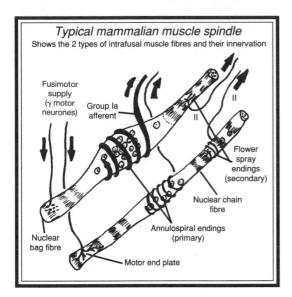

Typical mammalian muscle spindle
Shows the 2 types of intrafusal muscle fibres and their innervation

A muscle spindle consists of a bundle of small, thin, muscle fibres attached to a central sensory region containing the nerve ending of the afferent fibre. These intrafusal muscle fibres are innervated by gamma motor fibres. It is activity in these fibres which adjusts the sensitivity of the muscle spindle.

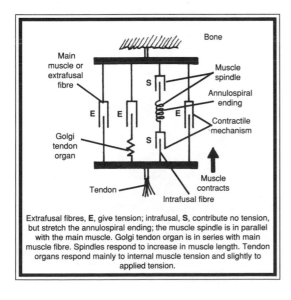

Extrafusal fibres, **E**, give tension; intrafusal, **S**, contribute no tension, but stretch the annulospiral ending; the muscle spindle is in parallel with the main muscle. Golgi tendon organ is in series with main muscle fibre. Spindles respond to increase in muscle length. Tendon organs respond mainly to internal muscle tension and slightly to applied tension.

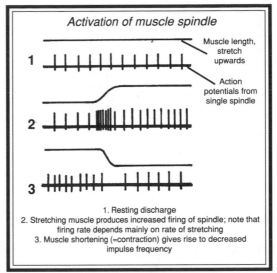

Activation of muscle spindle

1. Resting discharge
2. Stretching muscle produces increased firing of spindle; note that firing rate depends mainly on rate of stretching
3. Muscle shortening (=contraction) gives rise to decreased impulse frequency

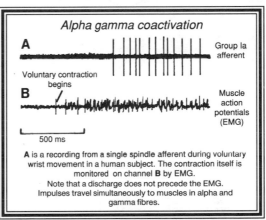

Alpha gamma coactivation

A is a recording from a single spindle afferent during voluntary wrist movement in a human subject. The contraction itself is monitored on channel **B** by EMG.
Note that a discharge does not precede the EMG.
Impulses travel simultaneously to muscles in alpha and gamma fibres.

Muscle spindles, in addition to producing mono-synaptic excitation of alpha motor neurones in their own muscles, at the same time cause a reflex inhibition of alpha motor neurones in their antagonist muscles. The mechanism for this is the activation of interneurones which inhibit the antagonist motor neurones. This system enables the shortening of synergistic muscles to be facilitated during the stretch reflex (reciprocal innervation, see Chapter 13).

A commonly quoted example of a stretch reflex is the knee jerk. This reflex is induced by tapping the tendon with a patella hammer. This tap produces a brief, quick stretch of the anterior muscle group in the thigh and, hence, the spindles contained in it.

To show the effects of the gamma motor system

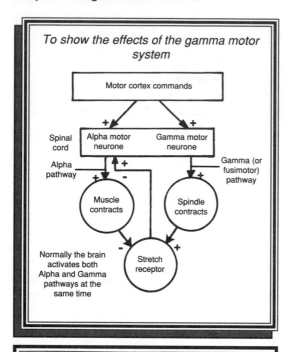

CONTROL OF MOVEMENT

It is often very difficult to distinguish between a voluntary movement and an involuntary one. For example, when learning a skill such as playing tennis, at first every movement has to be initiated voluntarily. However, as skill develops, the same movements become entirely automatic and the component parts of a stroke do not have to be thought out in detail. In other words, it looks as if the learned information about making the movement, has been relegated to some lower centre.

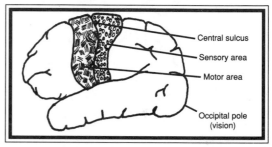

Golgi tendon organs

The tension in a muscle is signalled by Golgi tendon organs. These consist of branched nerve endings which are entwined within the tendinous fibres of a muscle. They activate group 1b sensory nerve fibres. When a tendon is stretched, the Golgi nerve endings are stimulated. While muscle spindles are in parallel with ordinary muscle fibres, Golgi tendon organs are in series with them. They do not give rise to a conscious sensation of force; their information is merely used in generating reflex responses, consisting largely of inhibition within the same muscle (autogenic inhibition).

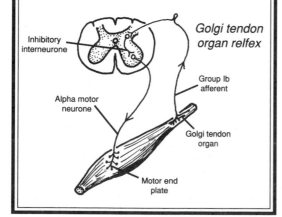

Golgi tendon organ relfex

Basal ganglia

The basal ganglia are in the forebrain beneath the cerebral hemispheres. They have connections with the cerebral cortex. Disease of or damage to the basal ganglia gives uncontrolled movement. A common example of this is Parkinson's disease which is characterised by muscular rigidity and tremor. There is considerabe difficulty in initiating voluntary movement. In this disease there is a deficit of the neuro-transmitter dopamine. This led to the development of treatment which involves giving a dopamine precursor (L-DOPA) to take the place of the deficient transmitter. The precursor is given because dopamine itself will not cross the blood brain barrier. The evidence suggests that the basal ganglia play a role in programming movements.

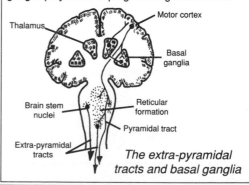

The extra-pyramidal tracts and basal ganglia

Cerebellum

This is a large organ which lies posterior to the medulla. The function of the cerebellum is the refinement and modification of movement. Its removal in an experimental animal does not abolish reflexes nor voluntary movements. However, all movements are grossly abnormal and have an error in rate of contraction, force of contraction and coordination. Lesions of the cerebellum in humans give rise to disturbances of balance, a staggering gait and generalised weakness. There is an impediment of speech. No sensory loss is involved and intellectual functions are unimpaired.

The cerebellum acts as an accessory control system which modifies the motor output of the cerebral cortex and brain stem. There is a detailed sensory input from receptors in skin, muscle and joints, the vestibular apparatus and the visual system. It appears that during any movement the cerebellum is acting as a feedback system which is continuously comparing the sensory input with the movement of muscles. The cerebellar output is then used to make corrections in the motor information which goes to the muscles concerned. The output of the cerebellum is inhibitory.

Reticular formation

The activity level of the nervous system changes in different circumstances. Levels vary from sleep to very alert. They are due to the degree of arousal produced in the brain as a result of changes of activity in the brain stem. The neurones responsible for this form the reticular formation. This is a diffusely organised network of cells which has projections to the cerebral cortex setting the level of consciousness. Some of the nuclei forming the reticular formation are actually part of the extra pyramidal system and have the function of maintaining balance and posture. Other parts of the reticular formation control respiration, heart rate and blood pressure. The inputs to the reticular formation come from throughout the body including most parts of the central nervous system. The outputs are also wide ranging . When the central part of the reticular formation is stimulated, the experimental animal or human subject becomes alert. This region is the ascending reticular activating system (ARAS). It has diffuse pathways ending in most of the cortex. Many of the pathways travel via synapses in the thalamus.

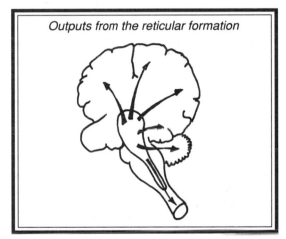

Outputs from the reticular formation

If electrodes are attached to the human scalp, electrical activity can be recorded. These potentials are known as the electroencephalogram (EEG). There is an empirical relationship between the waveforms of the EEG and the degree of alertness displayed by the subject. Abnormalities such as large brain tumours or epilepsy are associated with typical patterns in the EEG.

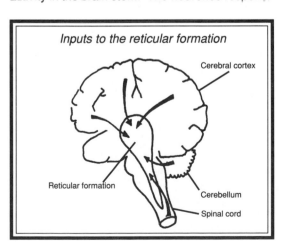

Inputs to the reticular formation

Cerebral cortex

Reticular formation

Cerebellum

Spinal cord

However, there is evidence which shows that many of the waveforms recorded from the scalp, are not generated by neurones in the brain substance. It seems fairly clear that they are artefacts generated by muscle movement (eye muscles, neck muscles).

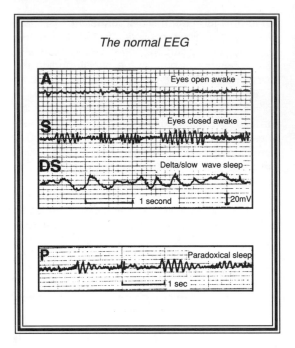

The normal EEG

Sleep

A large part of one's life is spent asleep; the physiological reason for this is quite unknown. The brain does not turn off during sleep, for instance a mother is roused instantly by the slightest cry of her child. This is one way in which sleep differs from the state of anaesthesia. Sleep is divided into two types: slow wave sleep and paradoxical sleep. When falling asleep, the alpha rhythm (10 Hz electrical waves) of the brain is replaced by slower and larger delta waves. A person is said then to be in slow wave sleep. At intervals this slow wave sleep is abolished by the so-called paradoxical sleep in which the pattern of the EEG looks much like that recorded from an awake subject. However, although the EEG response is one of awakeness the subject is in fact more deeply asleep because it is more difficult to arouse him. This paradoxical sleep is accompanied by lowered tone of the body muscles and is often associated with rapid eye movements. Hence it is also known as rapid eye movement sleep (REM sleep). It is during this time that we dream. Such segments of REM sleep usually continue for about 15 to 30 minutes each and comprise about a quarter of the time asleep.

Limbic system

The limbic system is in the forebrain. It comprises the limbic cortex (oldest phylogenetic part of the cerebral cortex), various sub-cortical areas and some parts of the hypothalamus, and gives rise to unstructured behavioural responses required for the preservation of life (and of the species), such as mating, reactions to hunger, thirst and danger. All these types of behaviour are modified by the neocortex, the most recently evolved region of the cortex, which includes the sensory, motor and association areas. The limbic system also plays a part in the emotional concomitants of sensation.

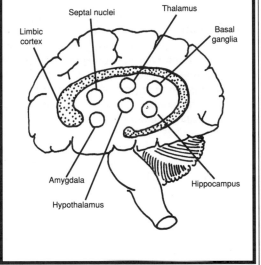

Learning and memory

A definition of learning is a change in response to a particular stimulus as a result of experience. Much research is currently being undertaken to delineate the cellular basis for learning and memory. Two types of memory occur: short term memory and long term memory. When learning a fact it is retained in short term memory for a period of several minutes to, possibly, an hour. This stored information may be disrupted by many factors such as a blow on the head, low body temperature, very deep anaesthesia and the administration of electroconvulsive therapy (ECT). Any of these procedures can give rise to a condition called retrograde amnesia. Memory of recent events disappears; the memory of earlier events usually being unimpaired.

Early memories are fairly resistant to this kind of disruption. Therefore the mechanisms involved in information storage in short term and long term memory are different. In short term memory reverberating circuits might pass impulses round closed chains of neurones and storage occurs in this kind of labile manner. After a while this transient form of storage is transferred into a more stable system, and probably involves some kind of structural change in neurones or synapses.

The storage of information presumably depends on systems which make use of altered transmission of information along nerve pathways. Such changes might reside either in cell bodies or synapses. Much work has been directed towards the demonstration of such plasticity in the brains of both higher and lower animal species. It is generally found that prolonged activation of a synapse, enhances its transmission properties. This alone would be a mechanism capable of explaining the known facts of memory and learning. Apart from the alteration of synaptic properties, new synapses might be formed or existing synapses might be removed. The administration of protein synthesis inhibitors tends to destroy the laying-down of long term memory and it has been proposed that the permanent changes in transmission characteristics depend on structural changes.

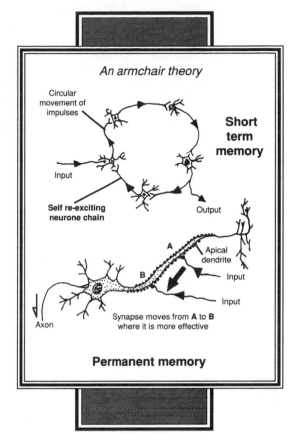

An armchair theory

Circular movement of impulses

Short term memory

Input

Output

Self re-exciting neurone chain

A — Apical dendrite

B

Input

Input

Axon

Synapse moves from **A** to **B** where it is more effective

Permanent memory

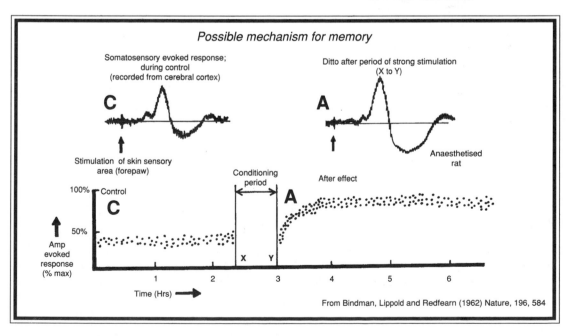

Possible mechanism for memory

Somatosensory evoked response; during control (recorded from cerebral cortex)

C

Ditto after period of strong stimulation (X to Y)

A

Anaesthetised rat

Stimulation of skin sensory area (forepaw)

Conditioning period

After effect

100% — Control

C

A

50%

Amp evoked response (% max)

X   Y

Time (Hrs) ➡

From Bindman, Lippold and Redfearn (1962) Nature, 196, 584

AUTONOMIC NERVOUS SYSTEM

The autonomic nervous system regulates most of the involuntary activities of the body. Some involuntary activities, however, do not involve the autonomic nervous system e.g. the knee jerk reflex. In addition the autonomic nervous system can exert a voluntary influence, such as the control of bladder emptying.

All efferent peripheral nerve fibres belong to the autonomic nervous system, except those efferents to striated muscles.

Autonomic pathways differ from somatic pathways in having two neurones between the central nervous system and the effector structure. The only exception is the innervation of the adrenal medulla.

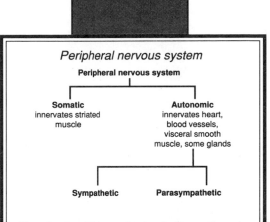

Peripheral nervous system

The glands which are innervated by autonomic nerves, include those associated with the digestive, respiratory and genito-urinary systems, and those of the head and skin. The visceral smooth muscles are found in the walls of the digestive, respiratory and genito-urinary tracts, the skin, the iris and ciliary body of the eye.

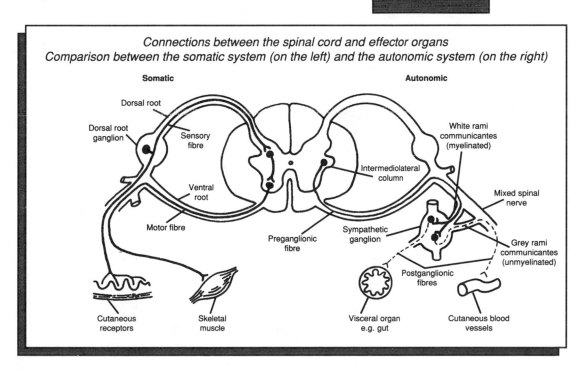

Connections between the spinal cord and effector organs
Comparison between the somatic system (on the left) and the autonomic system (on the right)

ANATOMY

Sympathetic nervous system (thoracolumbar outflow)

The sympathetic ganglia form a chain which is located immediately outside the vertebral column.

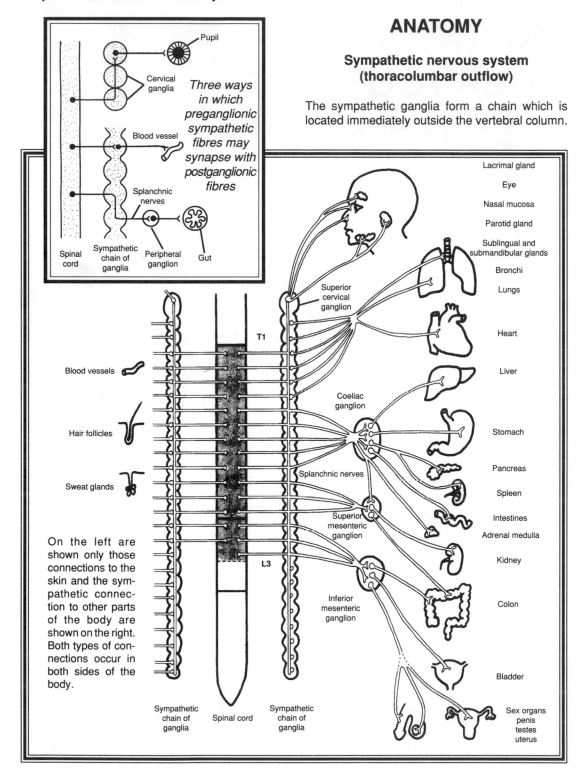

Three ways in which preganglionic sympathetic fibres may synapse with postganglionic fibres

Pupil

Cervical ganglia

Blood vessel

Splanchnic nerves

Spinal cord

Sympathetic chain of ganglia

Peripheral ganglion

Gut

Lacrimal gland

Eye

Nasal mucosa

Parotid gland

Sublingual and submandibular glands

Bronchi

Lungs

Heart

Liver

Stomach

Pancreas

Spleen

Intestines

Adrenal medulla

Kidney

Colon

Bladder

Sex organs
penis
testes
uterus

Superior cervical ganglion

T1

Coeliac ganglion

Splanchnic nerves

Superior mesenteric ganglion

Inferior mesenteric ganglion

L3

Blood vessels

Hair follicles

Sweat glands

On the left are shown only those connections to the skin and the sympathetic connection to other parts of the body are shown on the right. Both types of connections occur in both sides of the body.

Sympathetic chain of ganglia

Spinal cord

Sympathetic chain of ganglia

Parasympathetic nervous system
(craniosacral outflow)

The parasympathetic ganglia lie near to their effector organs or within their walls. These ganglia are often very small.

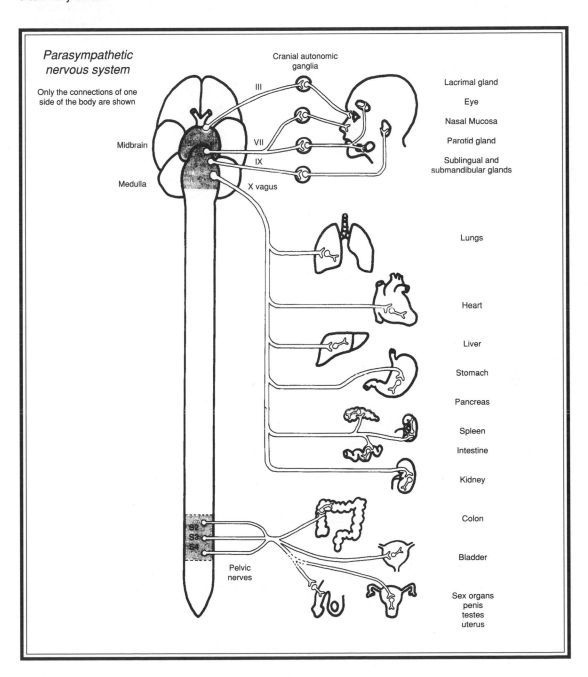

Parasympathetic nervous system

Only the connections of one side of the body are shown

Cranial autonomic ganglia

III

VII

IX

X vagus

Midbrain

Medulla

S2
S3
S4

Pelvic nerves

Lacrimal gland

Eye

Nasal Mucosa

Parotid gland

Sublingual and submandibular glands

Lungs

Heart

Liver

Stomach

Pancreas

Spleen

Intestine

Kidney

Colon

Bladder

Sex organs
penis
testes
uterus

PHARMACOLOGY

Autonomic nerve fibres release one of two transmitter substances, acetylcholine (cholinergic endings) or noradrenaline (adrenergic endings).

All autonomic preganglionic fibres and parasympathetic postganglionic neurones are cholinergic, while most of the sympathetic postganglionic neurones are adrenergic. The sympathetic nerves to the sweat glands and some nerves which produce vasodilation in muscle blood vessels, are cholinergic.

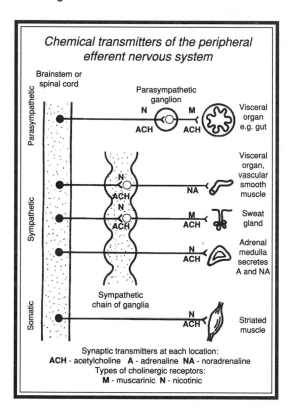

Chemical transmitters of the peripheral efferent nervous system

Synaptic transmitters at each location:
ACH - acetylcholine **A** - adrenaline **NA** - noradrenaline
Types of cholinergic receptors:
M - muscarinic **N** - nicotinic

The adrenal medulla

The medulla of the adrenal gland is innervated directly by cholinergic preganglionic sympathetic fibres. These fibres have synaptic connections with chromaffin cells which secrete adrenaline and noradrenaline as hormones (see Chapter 17). Hence the chromaffin cells may be considered as analogous to postganglionic neurones.

The postganglionic sympathetic transmitter, noradrenaline, produces both excitatory and inhibitory actions at its postsynaptic sites. These different responses in the effector organs are the result of the activation of two major types of adrenergic receptors, α and β receptors. The nature of the response will depend on which type of receptor is present in the target organ. On the whole α receptors mediate excitatory responses, while β receptors mediate inhibitory responses. However as usual there are exceptions. For example, activation of α receptors produces inhibition in the gastrointestinal tract, while the activation of β receptors stimulates the heart. Noradrenaline will strongly activate α receptors, but it only exerts a slight effect on β receptors. Adrenaline activates both types of receptors equally well.

The acetylcholine receptors of the autonomic nervous system are also of two types, muscarinic and nicotinic. Muscarinic receptors are activated by muscarine which elicits the same effects as acetylcholine in the target organs. This type of receptor is found in all the effector cells which are innervated by cholinergic, postganglionic neurones. Low doses of nicotine excite nicotinic receptors, while high doses block them. Nicotinic receptors are found in the postganglionic cells of the autonomic ganglia. They also occur in skeletal muscle at neuromuscular junctions, and in the chromaffin cells of the adrenal medulla.

The classification of autonomic receptors has led to the development of many drugs which can modify autonomic activity. These important drugs are employed to treat shock, high blood pressure and heart disease. Some drugs, mimetics, copy the actions of autonomic transmitters, while other drugs inhibit, reduce or reverse the responses that normally occur after autonomic nerve stimualtion (blockers).

FUNCTIONS

The sympathetic and parasympathetic nervous systems interact to regulate the body's autonomic functions. In general these systems have opposing actions, and the balance between their activities determines the observed effects on an organ.

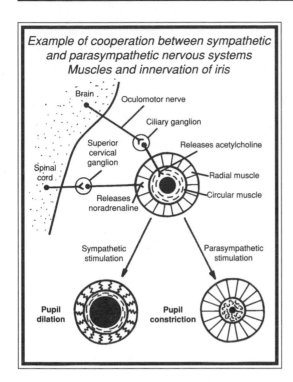

Example of cooperation between sympathetic and parasympathetic nervous systems Muscles and innervation of iris

Brain

Oculomotor nerve

Ciliary ganglion

Superior cervical ganglion

Releases acetylcholine

Spinal cord

Radial muscle

Circular muscle

Releases noradrenaline

Sympathetic stimulation

Parasympathetic stimulation

Pupil dilation

Pupil constriction

The autonomic nervous system maintains a basal level of activity. As a result, a continuous discharge in the sympathetic system keeps all the blood vessels constricted to about half their maximum diameter. Increased sympathetic discharge will further constrict the vessels, while decreased sympathetic discharge will dilate them.

When the sympathetic nervous system is strongly activated, a person becomes alert and able to perform strenuous physical activity and combat the three "Fs" (fright, flight and fight). Heart and respiratory rates increase; blood flow to skeletal muscles rises; sweating occurs and the pupils dilate. At the same time processes that interfere with muscular efficiency are inhibited, for example, digestion.

Activation of the sympathetic nervous system also causes the release of adrenaline and noradrenaline from the adrenal medulla. These hormones have almost the same effects as direct sympathetic stimulation, except that their influence persists a minute or two after the stimulation has finished, because they are removed slowly from the blood. There are some differences in the responses produced by adrenaline as opposed to noradrenaline.

Adrenaline is more powerful at increasing the metabolic rate and cardiac output.

The sympathetic nervous system often acts as a single unit to produce its responses to stress. However there are instances when restricted portions of the system are activated. For example, the sympathetic system helps regulate body temperature (see later). To this end it controls reflexes in which blood vessels, sweat glands and hair muscles are the effector organs.

Unlike the sympathetic system with its widespread discharge, most of the responses of the parasympathetic system are very specific, often only influencing single organs. The parasympathetic nervous system is responsible for the internal maintenance of the body, and is more active during recovery or rest. Hence increased parasympathetic activity reduces heart rate, enhances digestion and absorption of nutrients, brings about defaecation and emptying of the bladder.

HIGHER CENTRE CONTROL

Normal functioning of the autonomic nervous system is governed by regulatory centres located in the lower brainstem, hypothalamus and cerebral cortex.

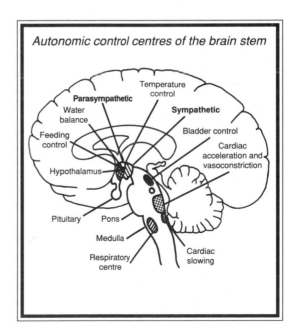

Autonomic control centres of the brain stem

Temperature control

Parasympathetic

Water balance

Sympathetic

Feeding control

Bladder control

Cardiac acceleration and vasoconstriction

Hypothalamus

Pituitary

Pons

Medulla

Respiratory centre

Cardiac slowing

Important medullary centres control heart rate, blood pressure and respiration. Other discrete groups of cells within the medulla and pons are concerned with different autonomic reflexes such as swallowing, intestinal movements, salivation, micturition and control of eye pupil size. Coordination between the autonomic and somatic nervous systems also occurs in the brain stem. An example of this coordination is the vomiting reflex. Vomiting requires the interaction of both skeletal and smooth muscles e.g. muscles of the abdominal wall, diaphragm, stomach and oesophagus.

The autonomic centres of the lower brainstem in turn receive an input from higher levels, particularly the hypothalamus. The hypothalamus is very important in the control of the internal environment, both via the autonomic nervous system and the endocrine system. In general the posterior hypothalamus influences the sympathetic nervous system, while the anterior areas control parasympathetic responses. There are also discrete regions in the hypothalamus which govern water balance (see Chapter 9), food intake (see Chapter 10) and body temperature. Both the sympathetic and the somatic nervous systems are involved in maintaining the body temperature at 37°C.

Almost all visceral responses can be modified by the activity of some area of the cerebral cortex. Autonomic and somatic effects have been shown to arise from the same cortical areas. For example, pupillary changes are elicited by stimulation of the eye field area, and at the same time conjugate eye movements are produced.

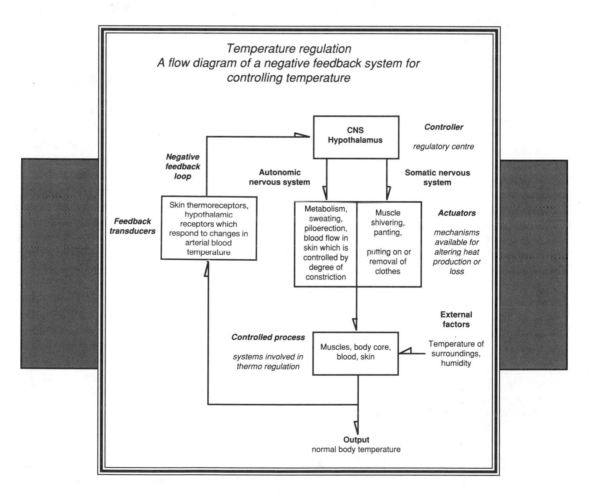

Temperature regulation
A flow diagram of a negative feedback system for controlling temperature

ENDOCRINES

A gland is an organ composed of secretory cells. There are two types of glands, exocrine and endocrine. The products of the exocrine glands are secreted into ducts e.g. sweat glands and mammary glands. The endocrines secrete straight into capillaries.

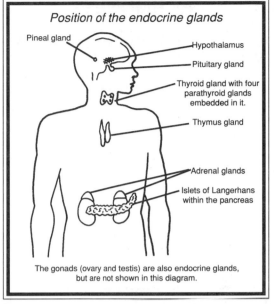

Position of the endocrine glands

Pineal gland

Hypothalamus

Pituitary gland

Thyroid gland with four parathyroid glands embedded in it.

Thymus gland

Adrenal glands

Islets of Langerhans within the pancreas

The gonads (ovary and testis) are also endocrine glands, but are not shown in this diagram.

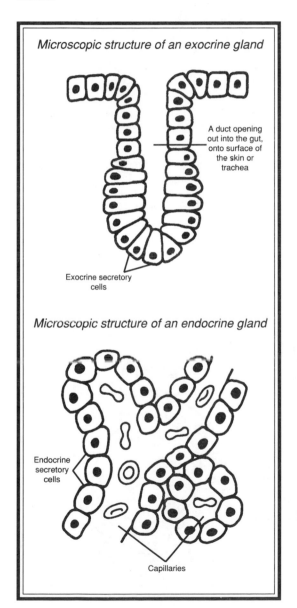

Microscopic structure of an exocrine gland

A duct opening out into the gut, onto surface of the skin or trachea

Exocrine secretory cells

Microscopic structure of an endocrine gland

Endocrine secretory cells

Capillaries

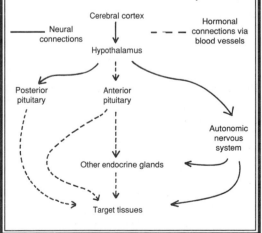

Relationship between endocrine and nervous systems

The endocrine system acts with the nervous system to control body function. The endocrine secretions (hormones) travel in the circulation, and cause effects throughout the body. The nervous system, on the other hand, generally produces rapid, discrete, localised effects.

The most important link between the two systems is the hypothalamus. It regulates the endocrine system through its influence on the pituitary gland and controls the autonomic nervous system.

Neural connections

Cerebral cortex

Hormonal connections via blood vessels

Hypothalamus

Posterior pituitary

Anterior pituitary

Autonomic nervous system

Other endocrine glands

Target tissues

HORMONES

Hormones are chemical messengers which can regulate the function of many organs. They are:-

- synthesised by cells that are usually grouped into an endocrine gland

- secreted directly into the blood stream

- secreted in minute amounts (typical hormone concentrations in the blood range from 10^{-6} to 10^{-12} moles/litre)

- carried by the blood to their sites of action

- able to speed up or slow down specific cellular reactions

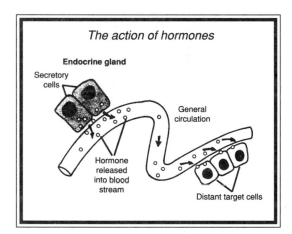

The action of hormones

(It should be realised that hormones are not enzymes or coenzymes, and neither do they provide metabolic energy.)

Hormones may be divided on the basis of their chemical structure, and the four types are proteins and peptides, steroids, iodothyronines and catecholamines.

Transport of hormones

Protein and peptide hormones circulate in an unbound form.

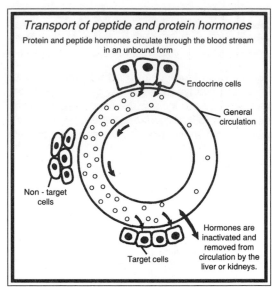

Transport of peptide and protein hormones

Protein and peptide hormones circulate through the blood stream in an unbound form

Steroid hormones usually circulate bound reversibly to plasma carrier proteins. This provides a means of protecting the hormone molecules from degradation in the liver or elimination via the kidneys. The protein bound hormones also represent an easily recruitable reserve.

Most hormones disappear from the circulation soon after they are secreted. This prevents their action being overly persistent.

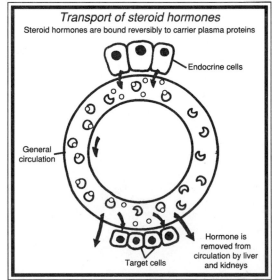

Transport of steroid hormones

Steroid hormones are bound reversibly to carrier plasma proteins

Hormone action

Protein, peptide and catecholamine hormones

These hormones are considered as the "primary messengers" and attach to specific receptor sites on the surface of the plasma membrane of their target cells. Binding of the hormone to the receptor produces cyclic 3'5'-adenosine monophosphate (cyclic AMP) which then acts as a "secondary messenger" within the target cell. The hormone itself does not pass through the cell membrane.

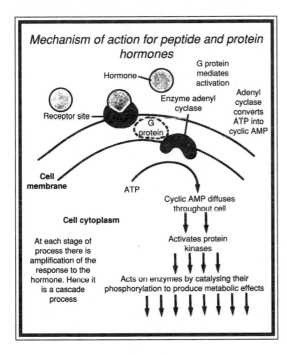

Mechanism of action for peptide and protein hormones

Steroid and iodothyronine hormones

When steroid hormones reach their target organs, they dissociate from their plasma binding proteins. They then diffuse through the plasma membrane and enter the target cell. Here the hormones bind to specific cytoplasmic receptor molecules. The hormone-receptor complex then passes into the nucleus where it binds reversibly to DNA. This initiates the synthesis of mRNA and consequently of proteins. Steroid hormones are, therefore, said to function as "gene-activators".

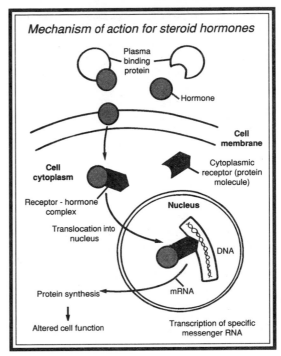

Mechanism of action for steroid hormones

The action of iodothyronine hormones is slightly different, in that the location of their receptor molecules is within the nucleus of the target cell rather than the cytoplasm.

Secretion of hormones

• External stimuli

Changing external conditions often affect the rate of hormone secretion. This allows the body to maintain a constant internal environment against the background of changing external environment.

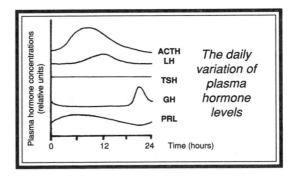

The daily variation of plasma hormone levels

Emotional and physical stress, for example, cause the release of cortisol and catecholamines. Secretion of gut hormones and insulin is stimulated by eating food.

• Internal rhythms

Hormone release is often rhythmical, and may follow daily, seasonal or yearly cycles These cyclical patterns are most obvious in the reproductive system e.g. the monthly ovarian cycle.

• Feedback control systems

Hormone secretion is mainly regulated by closed loop control systems. These keep the general activity of the target organs within well defined limits.

In a control system, the variable being controlled is measured, and the action performed is dependent upon a comparison between what is desired (input value) and what is achieved (output value). The output is "fedback" and influences the input to the system. The majority of the control systems found in the endocrine system have negative feedback. This means that the output of the system is subtracted from the input and the difference obtained serves as the signal to the controlled variable. However there are a few examples of positive feedback. Here the input and the fedback output are added to give the signal to the controlled variable. This results in an ever increasing output leading to a catastrophic situation. The best known example of a positive feedback control system is ovulation (Chapter18).

Examples of control systems found in the endocrine system

Basic negative feedback control system e.g. action of insulin on blood glucose levels

Input

Comparator — endocrine cells

Feedback signal concentration of metabolite or chemical substance

Hormone

Target cells — Effector

Output metabolite or chemical substance

Negative feedback control system through the anterior pituitary e.g. control of thryroxine secretion from thyroid

Anterior pituitary

Concentration of hormone in blood

Trophic hormone

Endocrine cells

Hormone

Target cells

Negative feedback control system involving the CNS

Hypothalamus

Releasing hormone

Anterior pituitary

Trophic hormone

Endocrine cells

Hormone

Target cells

Positive feedback control system

+ Anterior pituitary

FSH

Ovaries

Female sex hormones In most cases output increases until an event is triggered which switches off the system

Methods of investigating endocrine function

• Effects of gland removal

• Reintroduction of extracts

• Study of disorders in humans

PITUITARY GLAND

The pituitary gland, or hypophysis, secretes a variety of hormones whose common feature is a link with the hypothalamus. Either their secretion is controlled by hypothalamic hormones, or they are manufactured by hypothalamic cells. The pituitary is anatomically close to the hypothalamus.

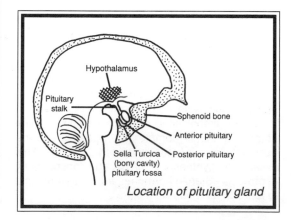

Hypothalamus

Pituitary stalk

Sphenoid bone

Anterior pituitary

Sella Turcica (bony cavity) pituitary fossa

Posterior pituitary

Location of pituitary gland

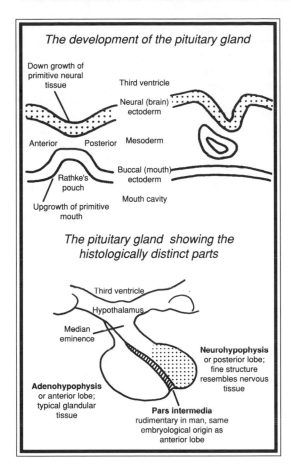

The development of the pituitary gland

Down growth of primitive neural tissue

Third ventricle

Neural (brain) ectoderm

Anterior   Posterior   Mesoderm

Rathke's pouch

Buccal (mouth) ectoderm

Upgrowth of primitive mouth

Mouth cavity

The pituitary gland showing the histologically distinct parts

Third ventricle

Hypothalamus

Median eminence

Neurohypophysis or posterior lobe; fine structure resembles nervous tissue

Adenohypophysis or anterior lobe; typical glandular tissue

Pars intermedia rudimentary in man, same embryological origin as anterior lobe

Anterior pituitary

The anterior pituitary secretes six peptide hormones: adrenocorticotrophic hormone (ACTH) or corticotrophin, luteinising hormone (LH), follicle-stimulating hormone (FSH), thyroid stimulating hormone (TSH), growth hormone (GH) and prolactin (PRL). All, except prolactin, are trophic hormones (they influence other glands).

The supply of arterial blood to the anterior pituitary is via portal vessels (they have a capillary bed at both ends). In this case the portal system arises from the capillaries of the median eminence within the hypothalamus. The importance of this arrangement is that the portal blood carries not only oxygen and nutrients, but also substances secreted by the hypothalamic neurones.

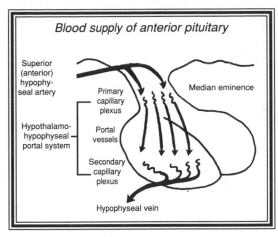

Blood supply of anterior pituitary

Superior (anterior) hypophyseal artery

Primary capillary plexus

Median eminence

Hypothalamo-hypophyseal portal system

Portal vessels

Secondary capillary plexus

Hypophyseal vein

Hypothalamus and hypothalamic hormones

The hypothalamus lies below the thalamus, and consists of clusters of neurone cell bodies, the hypothalamic nuclei.

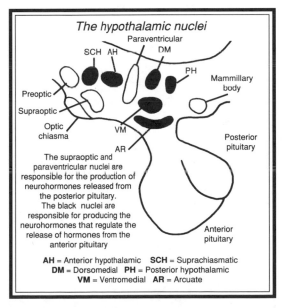

The hypothalamic nuclei

Paraventricular

SCH AH   DM

PH

Mammillary body

Preoptic

Supraoptic

Optic chiasma

VM

AR

Posterior pituitary

The supraoptic and paraventricular nuclei are responsible for the production of neurohormones released from the posterior pituitary. The black nuclei are responsible for producing the neurohormones that regulate the release of hormones from the anterior pituitary

Anterior pituitary

AH = Anterior hypothalamic **SCH** = Suprachiasmatic
DM = Dorsomedial **PH** = Posterior hypothalamic
VM = Ventromedial **AR** = Arcuate

The hypothalamic hormones are mainly small polypeptides. They are referred to as neurohormones as they are synthesised in the cell bodies of specialised hypothalamic neurones or neurosecretory cells. These neurohormones travel down the axons of the neurosecretory cells to be released into the capillaries of the median eminence.

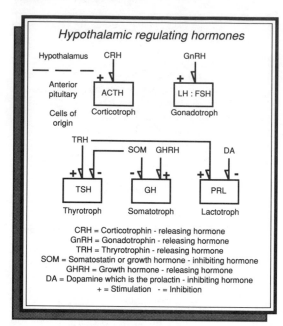

Hypothalamic regulating hormones

CRH = Corticotrophin - releasing hormone
GnRH = Gonadotrophin - releasing hormone
TRH = Thryrotrophin - releasing hormone
SOM = Somatostatin or growth hormone - inhibiting hormone
GHRH = Growth hormone - releasing hormone
DA = Dopamine which is the prolactin - inhibiting hormone
+ = Stimulation   - = Inhibition

The hypothalamus contains just a few nanograms of each releasing hormone. Only a fraction is secreted into the portal vessels at a time. However these minute quantities are sufficient to stimulate the anterior pituitary to release a thousand times that amount of trophic hormones. The trophic hormones in turn cause a thousand fold increase in secretion of hormones from their target glands. This is, therefore, a cascade process

The secretion of hypothalamic hormones is regulated by neural inputs from elsewhere in the brain as well as by feedback mechanisms. The CNS inputs include information about emotional state, external stimuli and biological rhythms.

Action of hypothalamic releasing or inhibitory hormones

Hypothalamic neurone

Median eminence

Releasing or inhibitory hormones

Hypophyseal portal vessels

Target cells in anterior pituitary

Pituitary hormone

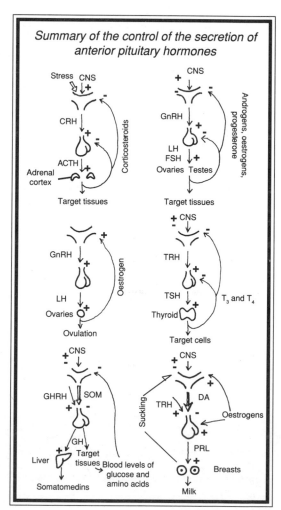

Summary of the control of the secretion of anterior pituitary hormones

Clinical disorders associated with pituitary hormones

Each of the pituitary hormones may be secreted either too much or too little for normal health. Usually hormone deficiencies are the result of the destruction or atrophy of the pituitary, while hormone excesses are caused by pituitary enlargement or tumours.

Excess or deficiency of TSH, FSH, LH or ACTH will lead to corresponding excesses or deficiencies of the hormones secreted from their target glands.

Prolactin

Prolactin deficiency produces no obvious ill effects, while excess leads to loss of menstruation and secretion of milk unassociated with pregnancy.

Growth hormone

Growth hormone has a wide range of effects including promotion of protein synthesis; increased breakdown of fats; raised blood sugar levels and increased red blood cell manufacture. It also causes the release of small peptides (somatomedins) from the liver and other tissues. These stimulate the formation of cartilage (important in determining skeletal growth).

Growth hormone deficiency is most significant in childhood and manifests itself as a failure of growth. An increased production of growth hormone in childhood produces gigantism, and can result in heights in excess of 2.5m. After puberty when the bones have stopped growing, an excess of growth hormone leads to acromegaly (enlarged feet and hands, and coarsening of the facial features).

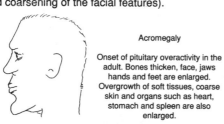

Acromegaly

Onset of pituitary overactivity in the adult. Bones thicken, face, jaws hands and feet are enlarged. Overgrowth of soft tissues, coarse skin and organs such as heart, stomach and spleen are also enlarged.

Posterior pituitary

The posterior pituitary is quite distinct from the anterior pituitary. They do not have a common blood supply. The posterior pituitary is composed of modified neuroglial cells (pituicytes), and the axon terminals of neurones (neurosecretory cells) whose cell bodies lie in the supraoptic and paraventricular nuclei of the hypothalamus. These neurosecretory cells release the peptide hormones, antidiuretic hormone (ADH or arginine vasopressin) and oxytocin, which have similar structures.

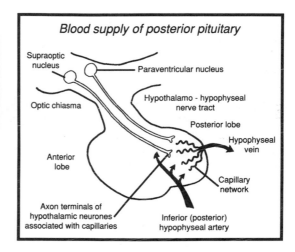

Blood supply of posterior pituitary

Large amounts of these hormones are stored in granules in the axon terminals within the posterior pituitary.

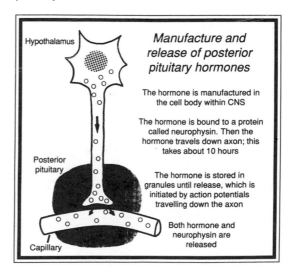

Manufacture and release of posterior pituitary hormones

The hormone is manufactured in the cell body within CNS

The hormone is bound to a protein called neurophysin. Then the hormone travels down axon; this takes about 10 hours

The hormone is stored in granules until release, which is initiated by action potentials travelling down the axon

Both hormone and neurophysin are released

The blood vessels of the posterior pituitary are outside the blood brain barrier, allowing hormones manufactured in the central nervous system to enter the general circulation.

Antidiuretic hormone (ADH)

The most important physiological action of ADH is to conserve water within the body; its principal target organ is the kidney. It acts by making the distal tubules and collecting ducts permeable to water; water is reabsorbed. Urine is more concentrated (see Chapter 9).

A failure of ADH secretion produces the disease, diabetes insipidus. Clinical features are the passage of large amounts (5-10 litres/day) of dilute urine, and a consequent excessive thirst.

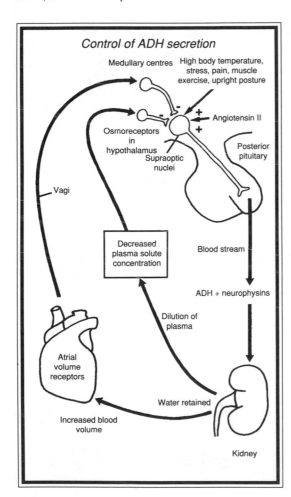

Oxytocin

Oxytocin is found in the posterior pituitary of both males and females, but both its known physiological effects occur only in females. It causes contraction of the myoepithelial cells of milk ducts in the breast, and milk is ejected. Oxytocin also stimulates contraction of the uterus. The sensitivity of the uterus and breasts to oxytocin increases during pregnancy and its effects are most pronounced once labour has actually started. This helps the fetus to be expelled. However, oxytocin is not an essential hormone in humans as labour can occur in its absence.

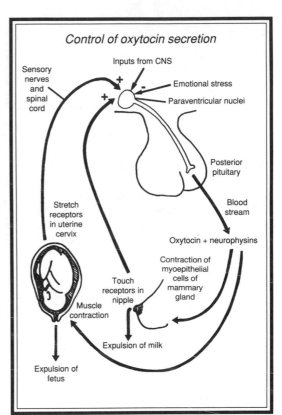

ADRENAL GLANDS

The adrenal glands consist of two distinct parts, an outer yellow cortex surrounding an inner darker coloured medulla. The two parts behave like separate glands; the cortex producing steroid hormones, while the medulla produces catecholamines.

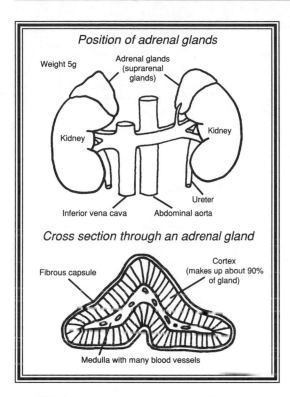

Position of adrenal glands

Weight 5g

Adrenal glands (suprarenal glands)

Kidney

Kidney

Inferior vena cava

Ureter

Abdominal aorta

Cross section through an adrenal gland

Fibrous capsule

Cortex (makes up about 90% of gland)

Medulla with many blood vessels

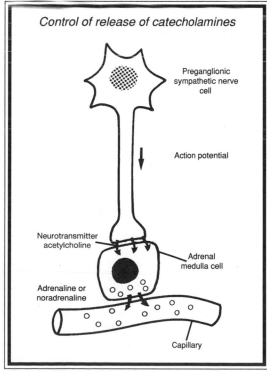

Control of release of catecholamines

Preganglionic sympathetic nerve cell

Action potential

Neurotransmitter acetylcholine

Adrenal medulla cell

Adrenaline or noradrenaline

Capillary

Adrenal medulla

The adrenal medulla contains large, ovoid, columnar cells arranged in clumps round blood vessels. The cells are packed with granules, which store the catecholamines, adrenaline and noradrenaline. Potassium dichromate stains the catecholamines a deep yellow brown, and hence the adrenal medulla cells are called chromaffin cells. There are two types of chromaffin cells, one type (the majority) secretes adrenaline, while the other secretes noradrenaline.

Release of catecholamines from the medulla is controlled by sympathetic nerves (splanchnic). Activity in them is invariably associated with activity in the rest of the sympathetic system. Catecholamines have a half-life of about five minutes in the circulation.

Medullary secretion results from emotional reactions, sexual excitement or stressful circumstances. In general the hormones of the adrenal medulla prepare the body for strong muscular activity and gear the body to give a fight or flight response (see Chapter 16).

Adrenal cortex

The adrenal cortex secretes three types of hormone: glucocorticoids which control carbohydrate metabolism and are involved in reactions to stress; mineralocorticoids which are concerned with salt and water metabolism; and sex hormones. All these hormones are steroids. Many different steroids have been isolated in the adrenal cortex, but only a few are actually secreted as hormones. The remainder are thought to be intermediates in the synthesis.

The principal hormones secreted from the adrenal cortex are:-

- Cortisol (hydrocortisone) - a glucocorticoid (20 mg/day)

- Aldosterone - a mineralocorticoid (100 μg/day)

- Dehydroepiandrosterone (DHA) - an androgen (20 mg/day)

Steroids

The basic steroid skeleton is composed of three six-carbon rings and one five-carbon ring.

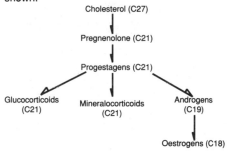

The different steroid hormones are produced by the addition to the basic skeleton of various side-groups. Above right is shown the basic structure of a mineralocorticoid or glucocorticoid. It has 21 carbon atoms, and so is referred to as a C21 steroid.

Steroid hormones are all derived, biosynthetically, from cholesterol. Cholesterol is either obtained from the diet, or it can be synthesised by the liver or, to a lesser extent, by the intestine. How the major classes of steroid hormones are related is shown.

Cholesterol (C27)

↓

Pregnenolone (C21)

↓

Progestagens (C21)

Glucocorticoids (C21)   Mineralocorticoids (C21)   Androgens (C19)

↓

Oestrogens (C18)

Histological structure of the adrenal gland

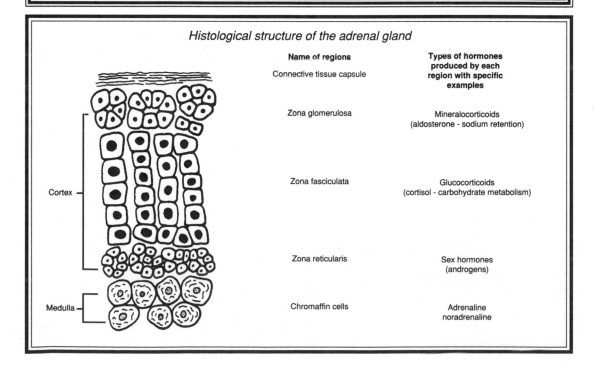

	Name of regions	Types of hormones produced by each region with specific examples
	Connective tissue capsule	
	Zona glomerulosa	Mineralocorticoids (aldosterone - sodium retention)
Cortex	Zona fasciculata	Glucocorticoids (cortisol - carbohydrate metabolism)
	Zona reticularis	Sex hormones (androgens)
Medulla	Chromaffin cells	Adrenaline noradrenaline

Glucocorticoids

Glucocorticoids exert many actions on a wide range of tissues. Their action on carbohydrate metabolism is concerned with the maintenance of the reserves of glycogen in the liver. This also involves the regulation of the amount of glucose in the blood, since glucose is stored as glycogen. Glucocorticoids raise blood sugar levels by stimulating the conversion of proteins into sugar (gluconeogenesis) in the liver. They also inhibit the uptake of glucose into peripheral tissues (e.g. muscle) by antagonising the action of insulin.

The secretion of cortisol and other glucocorticoids is controlled by the action of ACTH on the adrenal cortex. Secretion of ACTH is in turn under the control of the hypothalamic hormone, corticotrophin-releasing hormone (CRH). CRH release, and ultimately the secretion of cortisol, is influenced by a number of factors including:-

   • A diurnal rhythm which originates elsewhere in the brain, and is regulated by food intake rather than by light-dark or wake-sleep cycles. It results in diurnal variation in the amount of cortisol in the blood. Highest levels are found early in the morning and lowest (about half the maximum amount) between 10pm and midnight.

   • Negative feedback by the action of cortisol on both the hypothalamus and pituitary.

   • Physical and mental stress. This includes a variety of situations which are harmful, or potentially so, to the body, e.g. extremes of temperature, injury, infection, acute anxiety.

Mineralocorticoids

The role of mineralocorticoids is to maintain the balance between sodium and potassium ions in the body. Sodium ions are reabsorbed from sweat, saliva, urine and gastrointestinal contents. Aldosterone enters the epithelial cells of the distal convoluted tubules of the kidney and binds to cytoplasmic receptors. This activates ion transport mechanisns to reabsorb sodium ions in exchange for tubular secretion of potassium and hydrogen ions. Hence, mineralocorticoids also regulate water retention and extracellular fluid volume. The principal

mineralocorticoid is aldosterone(75% of the sodium retaining effect). In man the other 25% of the effect is caused by cortisol.

Control of aldosterone secretion

The secretion of aldosterone is increased by the factors shown. The most important is the renin-angiotensin system (see Chapter 9).

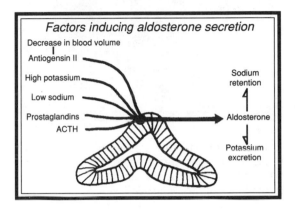

Factors inducing aldosterone secretion

Decrease in blood volume
Antiogensin II
High potassium
Low sodium
Prostaglandins
ACTH

Sodium retention
Aldosterone
Potassium excretion

THYROID GLAND

The thyroid gland secretes three hormones. Two of these, thyroxine (T_4) and tri-iodothyronine (T_3), stimulate oxidative metabolism in most cells of the body, and are necessary for growth and development. The third, calcitonin, lowers blood calcium levels.

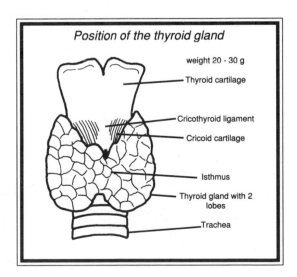

Position of the thyroid gland

weight 20 - 30 g
Thyroid cartilage
Cricothyroid ligament
Cricoid cartilage
Isthmus
Thyroid gland with 2 lobes
Trachea

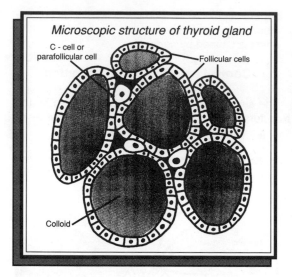

Microscopic structure of thyroid gland

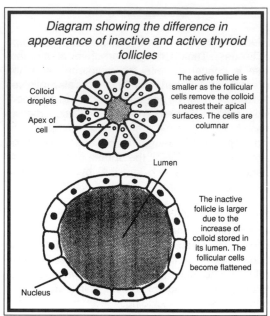

Diagram showing the difference in appearance of inactive and active thyroid follicles

The lobes of the thyroid are divided by connective tissue into lobules, each of which is composed of 20-40 spherical follicles. Each follicle comprises a single layer of cells surrounding a mass of semi-liquid, structureless substance known as colloid. The secretions of the thyroid cells are stored within this colloid. It contains about a month's supply of thyroid hormones. Interspersed among the thyroid follicles are C-cells or parafollicular cells, which secrete calcitonin.

Chemistry of thyroid hormones

The production of T_3 and T_4 requires iodine. Iodine is normally absorbed by the small intestine in the form of the iodide ion from food and drinking water.

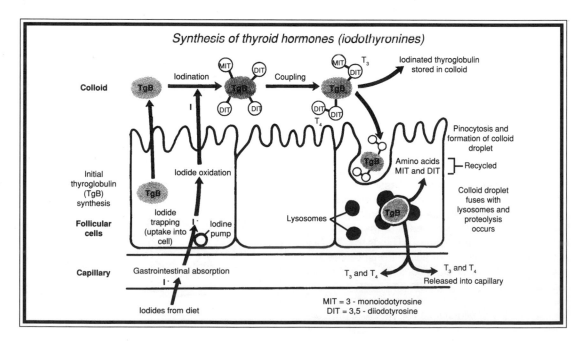

Synthesis of thyroid hormones (iodothyronines)

MIT = 3 - monoiodotyrosine
DIT = 3,5 - diiodotyrosine

The thyroid gland is very efficient at removing the iodide ions from blood. The follicular cells actively transport iodide across their basal membranes against both a concentration gradient and an electrical gradient. The concentration of iodide inside follicular cells is usually 25 to 50 times greater than the plasma concentration. This accumulation of iodide is referred to as iodide trapping.

Almost all of the released thyroid hormones are bound to specific plasma proteins. These binding proteins are thyroxine binding globulin (TBG), thyroxine binding prealbumin (TBPA) and albumin. However it is only the free unbound hormone in the circulation that is biologically active.

Actions of thyroid hormones

Thyroid hormones have several actions:-

- They increase the basal metabolic rate (see Chapter 10). This is associated with increased oxygen consumption and heat production (calorigenic effect) in all tissues except the brain, spleen and reproductive tissues. Hence they play a role in the maintenance of body temperature and adaptation to cold environments.

- They influence carbohydrate, protein and fat metabolism; important for normal growth and development.

- They are necessary for brain development of the fetus.

- They potentiate actions of other hormones such as catecholamines and insulin.

Control of thyroid hormone secretion

The secretion of thyroid hormones is controlled by the anterior pituitary hormone, TSH. It influences

---

Goitre

Enlargement of the thyroid gland is called a goitre, and arises when there is a lack of iodine in the diet. In certain areas of the world, there is very little iodine in the soil, and consequently in the vegetable crops. People in these areas may develop goitres.

---

practically all the steps in thyroid hormone production. TSH secretion is in turn governed by TRH (see earlier). External factors, such as cold and stress, modulate the hypothalamic output of TRH.

Diseases of the thyroid

Both an excess of thyroid hormones (hyperthyroidism) and a deficiency of thyroid hormones (hypothyroidism) can cause severe disorders. Thyroid malfunction is among the commonest of endocrine diseases, and is found in 2% of the adult population.

Hyperthyroidism

Hyperthyrodism results from an excessive production of thyroid hormones. The typical symptoms may include an acceleration of all physical and mental processes, tremor, nervousness and weight loss. The eyes of afflicted people appear to bulge out (exophthalmos).

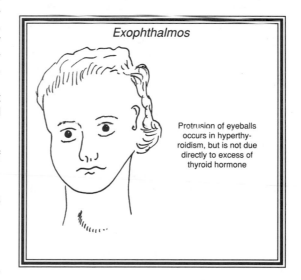

Exophthalmos

Protrusion of eyeballs occurs in hyperthyroidism, but is not due directly to excess of thyroid hormone

Hypothyroidism

Hypothyroidism is a deficiency of thyroid hormones. It results from either an impairment of TSH secretion, or a lack of ability to manufacture and secrete thyroid hormones. Patients with hypothyroidism suffer from a general slowing down both physically and mentally, weight gain, a dry coarse skin, loss of hair and weakness.

Calcitonin

Calcitonin is a polypeptide of 32 amino acids, which lowers blood calcium levels. Its site of action is on bone cells and it inhibits the movement of calcium from bone mineral to blood i.e. inhibits bone resorption. The secretion of calcitonin depends on the concentration of calcium ions in the blood. An increase in plasma calcium concentration will increase calcitonin secretion.

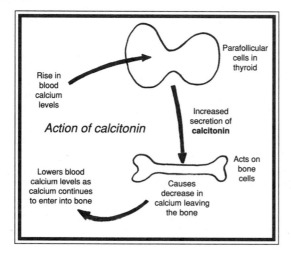

Action of calcitonin

PARATHYROID GLANDS

The parathyriod glands secrete parathormone which is important for calcium homeostasis. It is a polypeptide and regulates calcium plasma concentration by raising the calcium levels when appropriate. The opposite effect of lowering calcium plasma levels is brought about by calcitonin.

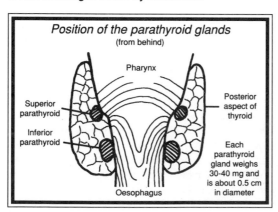

Position of the parathyroid glands
(from behind)

Calcium

Calcium plays an essential role in all basic body functions. It is required for:-

- formation and growth of bone and teeth
- blood clotting
- the action of many intracellular enzymes
- transmission of nerve impulses
- neurotransmitter release
- excitation-contraction coupling in muscle
- contraction of skeletal, cardiac and smooth muscle
- secretion from exocrine and endocrine glands
- mediation of hormonal effects
- growth of fetus during pregnancy
- milk production during lactation

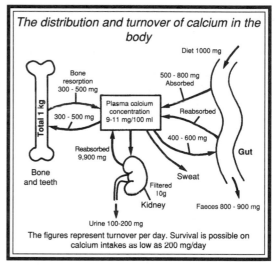

The distribution and turnover of calcium in the body

The figures represent turnover per day. Survival is possible on calcium intakes as low as 200 mg/day

The adult human contains 1-1.4 kg of calcium (1.5% of body weight). This is almost ten times the weight in the body of sodium and potassium combined. 99% of the calcium is found in bone which serves as a calcium store. The normal calcium level in blood plasma is 9-11 mg/100ml or 2.5 mmol/l. Just less than half of this is free calcium (unbound, ionised calcium), while the remainder circulates bound to either plasma proteins, primarily albumin, or anions such as citrate or lactate.

Calcium is obtained from the diet and absorbed through the small intestine. It is then either incorporated into bone or teeth, or else excreted again via urine or faeces. Usually very little is actually excreted in the urine unless the plasma concentration has risen above normal.

Phosphate

The calcium incorporated into bone and teeth is in the form of calcium phosphate. Consequently calcium homeostasis is linked to the regulation of phosphate; an increase in plasma calcium leads to a corresponding decrease in plasma phosphate and vice versa.

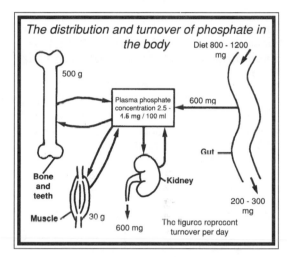

The distribution and turnover of phosphate in the body

All the body phosphorus occurs in compound form usually as phosphate combined with calcium or other inorganic cations, lipids or organic components. The term plasma phosphate refers to inorganic phosphate mainly as HPO_4^{2-}. Plasma phosphate concentrations in the adult vary between 2.5 and 4 mg/100ml (0.8-1.5 mmol/l). It does not appear to be critical for cellular function that plasma phosphate levels remain constant. Much greater variations from optimal can be tolerated than for calcium.

Phosphate, like calcium, plays an essential role in body function. It is important:-

• In the formation of bone and teeth

• As a component of many vital molecules e.g. nucleic acids, nucleotides, phosphoproteins, phospholipids in cell membranes

• For energy storage within cells in the form of ATP and GTP

• In the mediation of actions of many hormones i.e. conversion of ATP to cAMP

• In neural transmission

• For the activity of many enzymes

• As the major buffer system in urine

Most phosphate is obtained from the diet. Its absorption from the small intestine is linked to calcium absorption, and hence it is influenced by the same factors. Most important of these is the presence of the active form of vitamin D.

Bone

Bone is a protein matrix of inextensible collagen fibres onto which are deposited submicroscopic crystals of bone minerals. These crystals (hydroxyapatite crystals) are a mixture of calcium and phosphate, and provide bone with its rigidity. Between the collagen fibres is also a ground substance. This is mainly mucopolysaccharides and non-collagen proteins, and binds the collagen fibres together and so prevents compression. The heterogeneous composition of bone (i.e. bone is a material made of several different components) has the advantage that it combines the properties of each of its constituents.

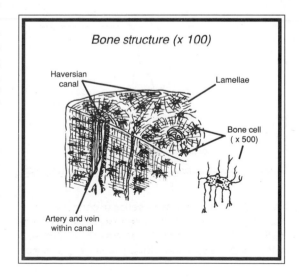

Bone structure (x 100)

Bone is not a static, fixed structure. Rather it is continually being broken down and built up. The process of breaking down bone is bone resorption, and involves both the release of calcium and phosphate by the dissolving of the hydroxyapatite crystals, and the removal of the collagen matrix. Bone resorption occurs on the surfaces of bony structures, and is brought about by the activity of large multinucleate cells called osteoclasts. The formation of bone, on the other hand, is carried out by uninucleate cells called osteoblasts. Osteoblasts synthesise and secrete collagen.

Vitamin D

Vitamin D refers to a group of steroid molecules which are produced by the action of ultra-violet light on various precursor sterols. It is also obtained from the diet. The form found in animals is known as vitamin D_3 and is synthesised by the skin in the presence of sunlight.

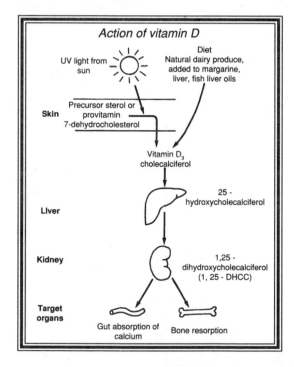

Action of vitamin D

Vitamin D has to undergo a series of conversions to reach its active form 1,25-dihydroxycholecalciferol (1,25-DHCC). 1,25-DHCC behaves like a hormone in that it travels in the blood stream to its target organs and only minute amounts are necessary to exert its effects. The main action of 1,25-DHCC is to promote the absorption of calcium from the gut. It is also involved in the resorption of bone mineral and hence the mobilisation of bone calcium.

A deficiency of vitamin D leads to rickets in children. This presents as a bowing of the legs, pain in the joints and stunting of growth. It results from low plasma calcium levels and consequent delayed calcification of bone.

Parathormone

Parathormone exerts its influence on calcium homeostasis at three different sites. The net result of its action is an elevation in the plasma concentration of calcium ions.

• The major site of action is bone. Parathormone promotes bone resorption, thereby releasing calcium and phosphate into the blood stream. This is probably mediated by osteoclasts, and requires 1,25-DHCC for maximum effect.

• Parathormone stimulates reabsorption of calcium from the distal tubules of the kidney, and depresses reabsorption of phosphate, causing more phosphate to be excreted. This is extremely important, and decreases plasma phosphate levels. As a result more calcium enters into the blood stream, so that the appropriate inverse relationship between calcium and phosphate plasma levels is maintained.

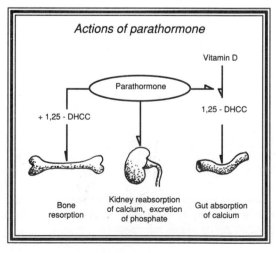

Actions of parathormone

• Parathormone has an indirect action on absorption of calcium from the small intestine, due to its stimulation of renal synthesis of 1,25-DHCC.

Secretion of parathormone

Parathormone secretion is controlled solely by the levels of free calcium ions in plasma, by means of a negative feedback system.

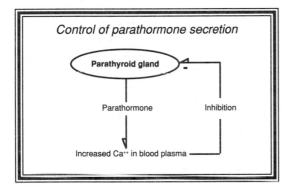

Control of parathormone secretion

Excess or lack of parathormone

The relatively common condition of an oversecretion of parathormone is hyperparathyroidism, usually due to a tumour. It results in bones becoming painful, weak and liable to bend or break, since too

Influence of other hormones on bone

Other hormones, apart from parathormone, calcitonin and 1,25-DHCC, are involved in bone formation and consequently calcium homeostasis.

• Growth hormone stimulates the production of somatomedins by the liver which in turn promote cartilage production and skeletal growth.

• Thyroid hormones cause normal skeletal growth and bone mineral exchange.

• Sex hormones stimulate bone growth and development. They are also involved in epiphyseal closure.

• Adrenal glucocorticoids in large amounts have a depressant effect on bone formation. They act by removing the bone protein matrix, and interfere with calcium absorption from the gut.

much calcium is being withdrawn. A lack of parathormone is hypoparathyroidism, and results in low levels of plasma calcium, causing increased neuromuscular activity and mental effects.

PANCREAS

The majority of cells in the pancreas have an exocrine function, and manufacture digestive enzymes. The other 1-2% have an endocrine function, and are located in the Islets of Langerhans surrounded by exocrine cells. The Islets are highly vascular areas, and in the human, number almost two million. Three types of cells have been identified in the Islets of Langerhans, α,β and δ cells. A fourth type of endocrine pancreatic cell are the PP cells which are scattered among the exocrine cells.

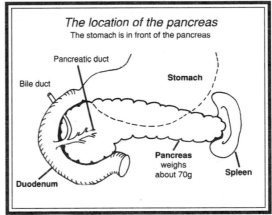

The location of the pancreas

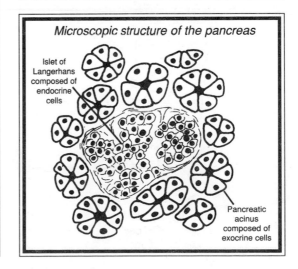

Microscopic structure of the pancreas

Human endocrine pancreatic cells			
Type of cell with alternative names	Proportion within islet	Hormone secreted	Function of hormone secreted
Alpha, α , A	20 - 30 %	Glucagon	Raises blood glucose levels
Beta , β , B	60 - 80 %	Insulin	Lowers blood glucose levels
Delta, δ , D	Up to 8 %	Somatostatin	Inhibits secretion of insulin, glucagon, growth hormone and TSH; also effects on gut and CNS
F, PP,D₁, type 4	Variable	Pancreatic polypeptide	Released after a meal but precise role unknown

Blood glucose levels

Glucose from the blood stream provides a major source of fuel for the cells of the body. Indeed it is the sole fuel source for the brain, and even very small decreases in blood glucose concentrations can cause harmful effects in the central nervous system. However , if the blood gucose levels rise too high, the kidneys become unable to reabsorb completely all the glucose from the renal tubules. This leads to loss of glucose from the body via the urine, and a consequent wastage of potential fuel. It is therefore essential to maintain blood glucose concentration within certain limits (80-100 mg/100ml or about 4.5 mmol/l).

The mechanism for maintaining a constant blood glucose level has to deal with the intermittent introduction of large quantities of glucose. Every time a carbohydrate-rich meal is eaten, glucose enters the blood stream via the intestinal capillaries. The utilisation of glucose can also be highly irregular, for example, the simultaneous activation of large groups of muscles requires greatly increased fuel supplies.

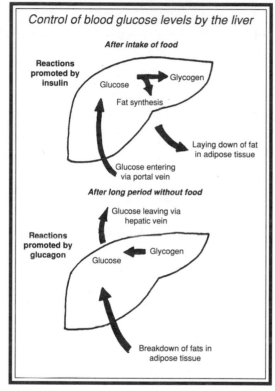

Control of blood glucose levels by the liver

Primarily the liver irons out variations in blood glucose levels. The liver is able to take glucose from the blood, and convert it into the polysaccharide, glycogen, which is stored in liver cells and provides a carbohydrate reserve. Since all the blood leaving the intestines enters the liver via the portal vein, the liver can remove the excess glucose following a meal. Hence the glucose concentration in the blood leaving the liver and joining the general circulation, is maintained within the required limits. When the amount of glucose entering the liver exceeds that which can be stored as glycogen, the liver converts the glucose into fat. If, conversely, the glucose levels of the blood entering the liver are

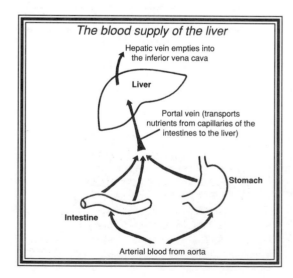

The blood supply of the liver

low, such as would occur in the morning before breakfast, the liver converts stored glycogen back into glucose. This glucose raises the blood glucose levels back to normal. The control of this hepatic activity involves the pancreatic hormones, insulin and glucagon.

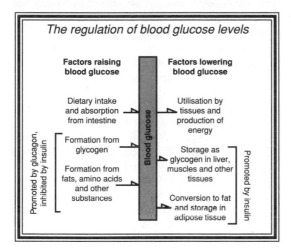

The regulation of blood glucose levels

Insulin influences protein and fat metabolism, by facilitating the transport of amino acids into liver and muscle cells, and inhibiting fat breakdown. The net effect of these actions is the synthesis of carbohydrates, proteins, lipids and nucleic acids. Hence insulin has a powerful anabolic action.

particular into muscle (including heart muscle) and adipose tissue. However, there are exceptions. Brain cells and red blood corpuscles are unresponsive to insulin. Glucose enters these cells by passively moving down a steep concentration gradient between blood plasma and the interior of the cell. Insulin also lowers blood glucose levels by stimulating the formation of glycogen from glucose in liver and muscle. At the same time it inhibits the break down of liver glycogen.

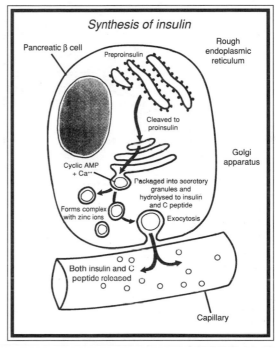

Synthesis of insulin

Insulin

The most important function of insulin is to lower blood glucose concentrations. To this end insulin acts on cell membranes in such a way as to allow the penetration of glucose into cells. Indeed insulin is required for entry of glucose into most cells, in

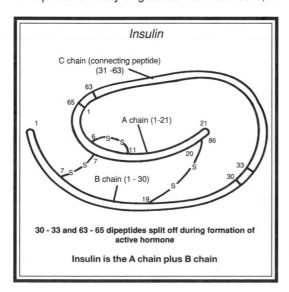

Insulin

30 - 33 and 63 - 65 dipeptides split off during formation of active hormone

Insulin is the A chain plus B chain

The secretion of insulin

The major stimulus to evoke the release of insulin is an increase in blood glucose concentration. Conversely a decrease in blood glucose concentration will cause decreased insulin secretion. This means that following a carbohydrate meal, there is a rise in insulin simultaneous with the rise in blood glucose level.

Insulin secretion is also controlled by many other factors, including the autonomic nervous system and other hormones. Obviously any agent capable of elevating blood glucose levels will stimulate insulin secretion.

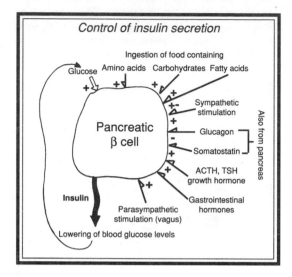

Glucagon is a polypeptide composed of 29 amino acids, and its basic effect is to raise blood glucose levels. This is achieved primarily by action on the liver. Glucagon increases the production of glucose from glycogen, and inhibits the synthesis of glycogen. Glucagon also mobilises glucose into the

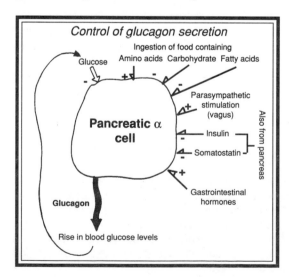

circulation by promoting its formation from amino acids and other noncarbohydrate sources (gluconeogenesis). The breakdown of fat is stimulated.

The main stimulus to elicit the release of glucagon is low blood glucose levels. Other factors, including hormones and nervous stimulation, also influence glucagon secretion. Many of these factors also influence insulin secretion.

Defects associated with pancreatic hormones

A lack of insulin causes the disease diabetes mellitus. Diabetes mellitus is characterised by the excretion of large quantities of sugar laden urine. This is the result of high blood glucose levels, and the kidneys being unable to reabsorb all the excess glucose. Normally no sugar should appear in the urine. The glucose that is not reabsorbed by the kidney tubules raises the osmotic pressure of the urine and consequently draws more water into the tubules. This leads to the production of large quantities of urine (polyuria). The patient then has an associated great thirst and will drink excessively (polydipsia). If the diabetes mellitus is left untreated death will ultimately ensue. It is in fact a leading cause of death in the United States and 1-2% of the world's population are afflicted. Nowadays the primary symptoms of diabetes can be controlled by insulin injections and/or dietary restrictions.

An excess of insulin leads to very low blood glucose levels (hypoglycaemia). The ultimate consequence of hypoglycaemia is a failure of the central nervous system as a result of the deficiency of glucose supplied to the brain cells. Initially the patient is just confused but then, if the problem remains untreated, he will collapse and go into a coma.

There is no known condition arising as the result of glucagon lack. Excess of glucagon causes the patient to be mildly diabetic, and occurs associated with stress, diabetes mellitus, chronic liver disease and chronic renal failure.

REPRODUCTION

REPRODUCTION

Since the life span of each individual within a species is finite, the continuation of that species relies on reproduction. Reproduction is the process by which an individual creates a new individual of the same species.

> The chemical carrier of hereditary material is deoxyribose nucleic acid (DNA). DNA is organised in the cell nucleus into a number of chromosomes. A short length of a chromosome, i.e. a particular sequence of bases of DNA, determines a precise trait or character of an individual. These portions of the DNA strand are referred to as genes and form discrete units of genetic material. Each gene is responsible for the manufacture of one specific protein.

In any human somatic cell (any cell apart from the ova and spermatozoa), there are 46 chromosomes which can be arranged into 23 pairs. Cells containing two sets of chromosomes like this are diploid. One set of chromosomes is derived from the mother, while the other is derived from the father. Twenty two of the pairs have members which look the same i.e. they are homologous. The twenty third pair are the sex chromosomes. In the female they are both X chromosomes, while the male has one X and a smaller Y chromosome.

A fundamental feature of sexual reproduction is that each individual receives two portions of genetic material. If the new individual was formed by the union of diploid cells from its parents, then every round of reproduction would result in a doubling of the chromosome number. This does not happen because a parent only transmits half his own chromosomes to his offspring. The gametes have only half the number of chromosomes. This number is the haploid number (N), and is 23 in humans. The process of fertilisation brings the gametes together to produce a zygote with 46 chromosomes. This then divides mitotically.

The gametes, ova and spermatozoa, are derived from primordial germ cells, that are present very early in embryonic development, and possess pseudopodia with which they migrate into the developing gonads (ovaries or testes). During this migration, the germ cells multiply by mitosis from about 100 early on, to about 5000 in the gonads. Once the germ cells have colonised the gonads, they continue to multiply and ultimately develop into oogonia or spermatogonia according to whether they are female or male. These cells are the precursor cells of the gametes. Before birth all the oogonia have become transformed by mitotic divisions into oocytes. The numbers of oocytes within the ovary cannot increase any further from this time. The spermatogonia, in contrast, persist in the testes throughout the life of the male, and are capable of multiplication. The spermatogonia divide mitotically to form spermatocytes. Both the spermatocytes and the oocytes are diploid cells with 46 chromosomes. These cells then undergo a special form of cell division called meiosis to produce the gametes. Meiosis (reduction division) produces by

A human karyotype

This is an ordered display of an individual's chromosomes taken from a typical somatic cell. The karyotype here is from a male as 23rd pair of chromosomes, the sex chromosomes, consists of an X chromosome and a Y chromosome

1 2 3 4 5
6 7 8 9 10 11 12
13 14 15 16 17 18
19 20 21 22
X Y

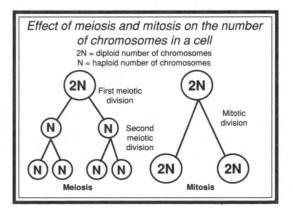

Effect of meiosis and mitosis on the number of chromosomes in a cell

2N = diploid number of chromosomes
N = haploid number of chromosomes

two successive cell divisions, four daughter cells each of which have half the chromosome number of the parent cell.

The result of combining genetic material from two parents is that the offspring is not identical to either parent, or to any of its siblings. This arises because the parental chromosomes are independently assorted and divided into half during meiosis. Another property of meiosis that introduces variability is that genetic information can be exchanged between pairs of homologous chromosomes (recombination). Finally, futher variation is brought about by the random combination of an ovum and a spermatozoon.

As all male somatic cells carry both an X and a Y chromosome, and the spermatozoa have only half the total set of chromosomes, each spermatozoon has either an X or a Y chromosome. However, all the ova have X chromosomes, since female somatic cells have two X chromosomes. When a Y spermatozoon fertilises an ovum, the offspring will be male. Alternatively when an X spermatozoon fertilises an ovum, the result will be female.

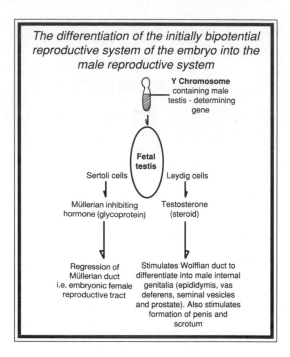

The differentiation of the initially bipotential reproductive system of the embryo into the male reproductive system

The continued differentiation of an embryo into a normal male now relies on secretion of two hormones from the fetal testes. Testosterone is secreted from Leydig cells lying between the seminiferous tubules, and Müllerian inhibiting hormone is secreted by the Sertoli cells.

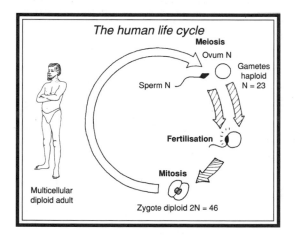

The human life cycle

The initial formation of the gonads in the embryo is identical for both sexes. It is the presence of the Y chromosome that causes the transformation of the gonad into a testis. This happens towards the end of the second month of development in utero. In the absence of a Y chromosome the gonads develop into ovaries.

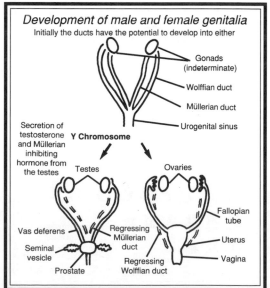

Development of male and female genitalia
Initially the ducts have the potential to develop into either

Like the gonads, the precursors of the internal genitalia have the potential to become either male or female.

Initially the embryo has two parallel duct systems, the Wolffian ducts and the Müllerian ducts. Testosterone stimulates the Wolffian ducts to differentiate into the male reproductive tract. Meanwhile Müllerian inhibiting hormone causes the degeneration of the Müllerian ducts. If these hormones are not present, the Wolffian ducts regress and the Müllerian ducts develop into the oviducts, uterus and upper vagina. Hence in the absence of hormonal influence, regardless of whether the ovary is present or not, the female reproductive tract develops.

When the fetus is about eight months old the testes migrate over the pelvis into the scrotal sack. In this position the temperature of the testes are several degrees lower than that of the abdomen interior. It appears that this lower temperature is required for the normal production of spermatozoa. The ovaries on the other hand remain close to their original position in the abdominal cavity.

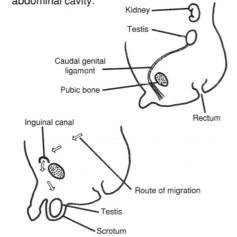

GROWTH

By the time of birth, the anatomy of the reproductive system has differentiated to the stage at which it remains until puberty. Between birth and adolescence the individual grows in a continuous fashion.

However different parts of the body grow at different rates: the growth of the reproductive organs proceeds at a considerably slower rate than the growth of the muscles, digestive system and bones.

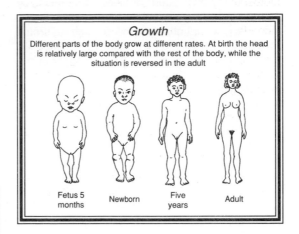

Growth

Different parts of the body grow at different rates. At birth the head is relatively large compared with the rest of the body, while the situation is reversed in the adult

| Fetus 5 months | Newborn | Five years | Adult |

The testes grow slowly and continuously during early life, but they secrete only minimal amounts of hormones. Likewise during childhood the ovaries are quiescent and change little. Nevertheless the ovaries at the time of birth contain all the oocytes they will ever have, about two million in total. During childhood the majority of the oocytes will degenerate and be lost. By the time puberty is reached only a fraction of the oocytes are left, about 400,000.

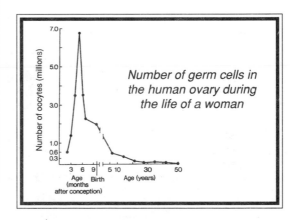

Number of germ cells in the human ovary during the life of a woman

Puberty is the time at which an immature individual becomes capable of reproducing. In girls the most obvious sign that puberty is occurring is the first menstrual bleeding, the menarche. The equivalent situation in boys is the ability to ejaculate semen.

Prior to these events a number of other physical changes are initiated which lead to the development of the secondary sexual characteristics. Firstly there is a marked increase in rate of growth, the adolescent "growth spurt". This is accompanied by changes in body composition. After this the various secondary sexual characteristics, such as breast development, genital growth, adult hair distribution and voice changes, occur in a definite time sequence. The sequence may start at very different chronological ages in different individuals but it always occurs in the same order.

The initiation of puberty corresponds to an increase in secretion of gonadotrophins from the anterior pituitary. The gonadotrophins in turn cause the release of steroid hormones from the gonads. The testes produce testosterone which induces the male secondary sexual characteristics.

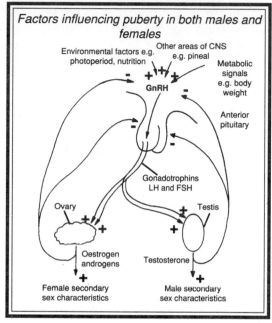

Factors influencing puberty in both males and females

Rate of growth for boys and girls showing pubertal growth spurt

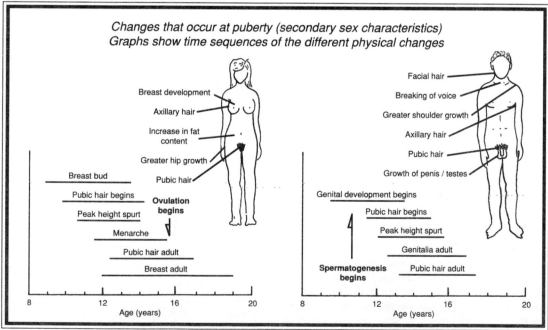

Changes that occur at puberty (secondary sex characteristics)
Graphs show time sequences of the different physical changes

The ovaries on the other hand start to produce oestrogen which causes breast enlargement and development of the lining of the uterus. The ovaries also begin secreting androgens which are responsible for the female adult hair growth patterns.

Gonadotrophin secretion is controlled by gonadotrophin releasing hormone (GnRH) which originates in the hypothalamus. The exact cause for the increased secretion at the onset of puberty is unknown. However influences from elsewhere in the CNS, in particular the pineal gland, are likely to be involved. The menarche is also closely associated with attainment of a body weight of 47 kg.

MALE REPRODUCTIVE SYSTEM

The male reproductive system manufactures and stores the male gametes, and then delivers them to the neighbourhood of the female gametes. The male gametes or spermatozoa are manufactured in the testes and stored in the epididymis. The penis provides a delivery mechanism, while the accessory sex glands contribute various secretions which aid the transport of the spermatozoa.

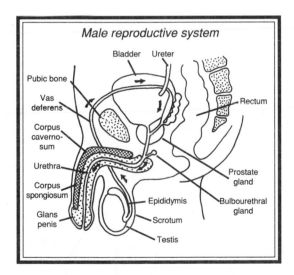

Male reproductive system

The testis has two functions, the production of steroid hormones and spermatogenesis. These are carried out in anatomically distinct compartments. The endocrine function is performed by cells lying between the seminiferous tubules, while

spermatogenesis, the manufacture of spermatozoa, takes place within the tubules. Indeed the composition of the fluid within the tubules is markedly different from that of the interstitial spaces between the tubules; these compartments are separated by a physical barrier, the blood-testis barrier.

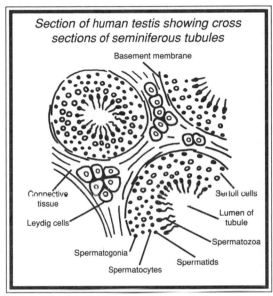

Section of human testis showing cross sections of seminiferous tubules

Endocrine role of testis

In response to stimulation by LH from the anterior pituitary, the Leydig cells of the testis produce testosterone (4-10mg/day).

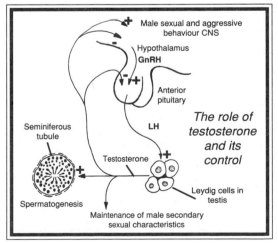

The role of testosterone and its control

Testosterone is an androgen i.e. a C19 steroid with a masculinising or virilising effect. Hence the action of testosterone is to develop and maintain masculine traits. It can pass through the blood-testis barrier. This means that since the Leydig cells are very close to the seminiferous tubules, there are high concentrations of testosterone around the developing spermatozoa. This is vital for spermatogenesis. Testosterone also influences the CNS and is responsible for the aggressive or positive attitude of males.

Sertoli cells

The seminiferous tubules contain only two types of cells, Sertoli cells and germ cells. A Sertoli cell has a multilobed nucleus and a large amount of cytoplasm that extends from the base of the seminiferous tubule to the central lumen. In the mature testis the Sertoli cells do not divide. Between neighbouring Sertoli cells are tight junctions that help to form the blood-testis barrier. The Sertoli cells produce the tubular fluid. They also play a role in spermatogenesis and mediate the actions of FSH and testosterone. Indeed the developing germ cells are very closely associated with the Sertoli cells which provide them with mechanical support and nourishment.

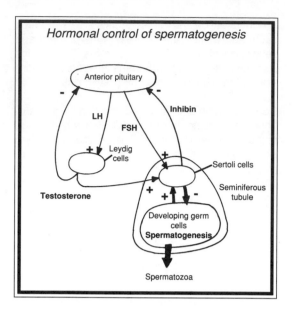

Hormonal control of spermatogenesis

Spermatogenesis

Spermatogenesis is the process of producing mature spermatozoa. From birth the division of the germ cells is held in abeyance until puberty. At this point spermatogenesis begins and the original germ cells become known as spermatogonia.

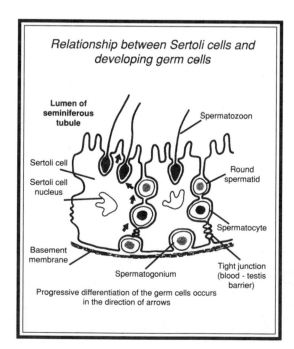

Relationship between Sertoli cells and developing germ cells

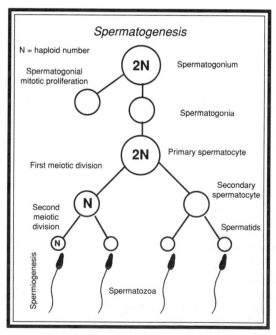

Spermatogenesis

The spermatogonia divide mitotically to give more spermatogonia. Some of these mature and increase in size to form primary spermatocytes which divide meiotically. The first division of meiosis forms secondary spermatocytes and the second forms spermatids. Initially spermatids are spherical with nuclei 5-6µm across. Each successive division during spermatogenesis moves the germ cells towards the centre of the seminiferous tubule so that the spermatids are closest to the lumen. The final step of spermatogenesis is spermiogenesis in which spermatids are converted, without further cell division, into mature spermatozoa.

These stages are equally spaced in time, which means that new sets of spermatozoa arrive in the tubule lumen every 16 days.

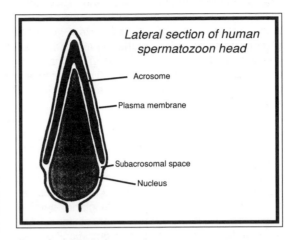

Lateral section of human spermatozoon head

- Acrosome
- Plasma membrane
- Subacrosomal space
- Nucleus

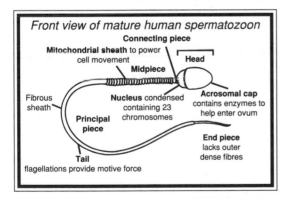

Spermiogenesis

Early spermatid

Acrosome forming over nucleus

Late spermatid

Nucleus and acrosome take characteristic shape

Nucleus

Tail formation beginning

Mitochondria

Mid piece forming, containing mitochondria for energy generation

Loss of cytoplasm, discarded as the residual body contains ribosomes and golgi membranes which are not needed by spermatozoon

Front view of mature human spermatozoon

Connecting piece

Mitochondrial sheath to power cell movement

Midpiece

Head

Fibrous sheath

Nucleus condensed containing 23 chromosomes

Principal piece

Acrosomal cap contains enzymes to help enter ovum

End piece lacks outer dense fibres

Tail flagellations provide motive force

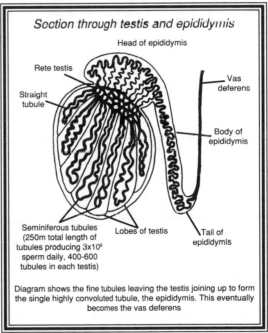

Section through testis and epididymis

Head of epididymis

Rete testis

Straight tubule

Vas deferens

Body of epididymis

Seminiferous tubules (250m total length of tubules producing 3×10^8 sperm daily, 400-600 tubules in each testis)

Lobes of testis

Tail of epididymis

Diagram shows the fine tubules leaving the testis joining up to form the single highly convoluted tubule, the epididymis. This eventually becomes the vas deferens

In man the time for a complete cycle of spermatogenesis, from the first division of spermatogonia to the release of spermatozoa, is 64 days. Along any radius of a seminiferous tubule, there are germ cells at four different stages of the cycle.

When the spermatids have aquired their tails and become spermatozoa, they are released into fluid in the centre of the tubule. They are still immature and incapable of movement, so they are carried by the fluid towards the epididymis. The epididymis is a highly convoluted tube measuring 5-7µm across, and it takes 12 days for the spermatozoa to pass through. The spermatozoa are stored here until release. This overcomes the problem of continuous production of spermatozoa which can only be released intermittently by ejaculation. During this storage period the spermatozoa become mature and motile. Various substances such as glycoproteins are added to the spermatozoan suspension, while water is removed. This change in the fluid composition leads to a 100 fold increase in spermatozoa concentration. There are now 50×10^8 sperm/ml.

The spermatozoa are then transported along the vas deferens by muscular contractions of its walls, and further fluids are added to make the semen which will be injected into the female tract. These additions, the seminal fluid, come from the accessory sex glands i.e. the prostate gland, seminal vesicles and bulbo-urethral (Cowper's) glands. The seminal fluid contains high concentrations of fructose (1.5mg/ml), and hence provides nutrients for the spermatozoa. It also provides protection.

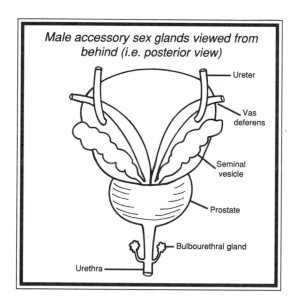

Male accessory sex glands viewed from behind (i.e. posterior view)

Ureter

Vas deferens

Seminal vesicle

Prostate

Bulbourethral gland

Urethra

Penis

The final path of the spermatozoa in the male reproductive tract is along the urethra to be expelled from the penis during coitus. The penis provides the means of depositing semen in the female tract. In order to accomplish this, the penis needs to be rigid (erect) so that it can enter the vagina.

This is achieved by the erectile tissues of the penis, two corpora cavernosa (lateral bodies) and the central corpus spongiosum. They consist of cavernous venous sinuses bound by fibroelastic tissue. During sexual arousal, the arteries feeding the sinuses become dilated. Consequently there is a great increase in the blood volume flowing into the sinuses and they expand. Meanwhile the veins which would normally drain the sinuses are compressed and prevent the blood leaving. Thus the sinuses rapidly become turgid, and the penis stiffens.

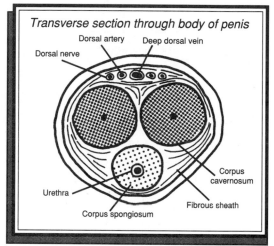

Transverse section through body of penis

Dorsal artery   Deep dorsal vein

Dorsal nerve

Corpus cavernosum

Urethra

Fibrous sheath

Corpus spongiosum

Movement of the semen from the vas deferens into the urethra is achieved by smooth muscle contraction. Further rapid rhythmic contractions of the vas deferens and the muscles around the penis expel the semen from the penis (ejaculation).

Female reproductive system

The female reproductive system is based on the ovaries where female germ cells are matured and stored. The system must be able to receive male gametes and bring about their unison with female gametes or ova (fertilisation). This is achieved by the deposition of spermatozoa in the vagina during copulation. The spermatozoa can then swim through the female reproductive tract to meet with ova that are released from the ovaries into the Fallopian tubes. Finally the fertilised ova require a safe site in which to develop into an embryo and fetus. This occurs in the uterus. When the fetus has reached sufficient maturity to survive as a separate individual, it is expelled from the uterus via the vagina (it is born).

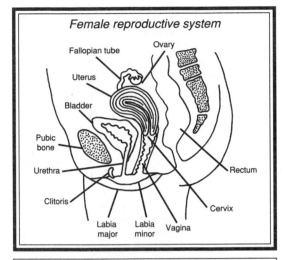

Female reproductive system

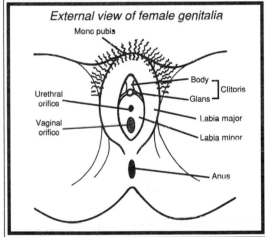

External view of female genitalia

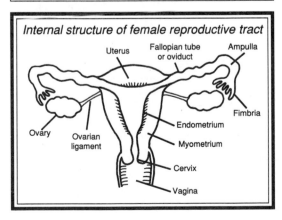

Internal structure of female reproductive tract

The production of female gametes, oogenesis, follows a very similar pattern to spermatogenesis. In both cases diploid primordial germ cells multiply mitotically and then progress through a number of divisions to become haploid gametes. However there are some important differences related to the fact that females only produce relatively few fully mature gametes, as opposed to the continuous massive output of spermatozoa from the testes.

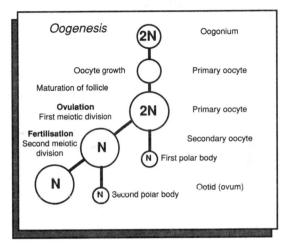

In the female the proliferation of germ cells producing oogonia, and the mitotic division of the oogonia to give oocytes are completed before birth.

The primary oocytes become surrounded by flattened ovarian cells to form primordial follicles. Thereafter the primary oocytes proceed in one of three ways. The vast majority die and are lost. Many of them have their development arrested and can remain in primordial follicles for up to 50 years. A few of the primordial follicles start to grow. Firstly there is an increase in the size of the primary oocyte, owing to an expansion of the cell cytoplasm. The primary oocytes may increase their diameters from less than $30\mu m$ to over $100\mu m$. The cells surrounding the oocyte change to a cuboidal shape and multiply. These cells are now granulosa cells, and they secrete a glycoprotein material which forms the zona pellucida immediately encompassing the oocyte. A layer of spindle-shaped ovarian stromal cells form around the outside of the granulosa cells. These constitute the thecal cells, and the follicle has reached the stage referred to as preantral.

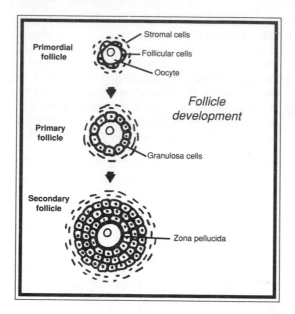

Follicle development

The growth of the follicle until it reaches the preantral stage, is independent of hormonal influences.

However at this point the granulosa cells develop receptors for FSH and the thecal cells receptors for LH. These gonadotrophins are necessary for any further growth of the follicle. Without the appropriate hormonal stimuli, the follicles undergo degeneration and the oocyte dies. This is known as atresia. Atresia is in fact the fate of the majority of follicles, and indeed all follicles which start to grow before puberty become atretic.

The effects of gonadotrophins is to convert the preantral follicles into antral follicles (vesicular or Graafian follicles). The granulosa and thecal cells multiply and the volume of the follicle increases. The thecal cells divide into two layers, the glandular theca interna and the fibrous theca externa. Fluid collects in spaces between the granulosa cells, and eventually these merge to form the fluid-filled antrum. The primary oocyte becomes surrounded by a densely packed mass of granulosa cells, the cumulus oophorus.

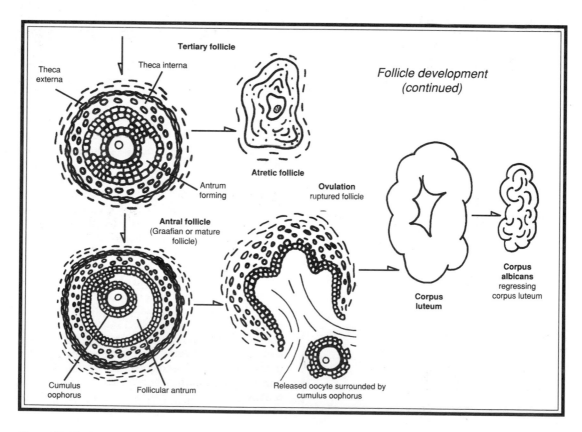

Follicle development (continued)

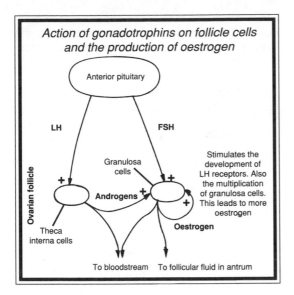

Action of gonadotrophins on follicle cells and the production of oestrogen

During formation of the antrum, the theca interna cells start to produce androgens in response to stimulation with LH. In the presence of FSH, these androgens are converted by the granulosa cells to oestrogens. The oestrogens either accumulate in the follicular fluid of the antrum or are released into the bloodstream. They can then bind to receptors within the granulosa cells, and this stimulates the production of more granulosa cells. This in turn stimulates the manufacture of yet larger amounts of oestrogens. If the oestrogen concentration in the blood reaches a sufficiently high level for a certain length of time, it can exert a positive feedback effect on the release of gonadotrophins from the pituitary. Hence, there is a dramatic increase in the secretion of LH, sometimes referred to as the midcycle LH surge.

Oestrogen has another effect on granulosa cells. Together with FSH, it stimulates the appearance of LH-binding sites, or receptors, on the granulosa cells. The presence of these granulosa cell LH receptors enables a follicle to respond to the LH surge mentioned above. Follicles that have not developed LH receptors are unable to respond to the LH surge and undergo atresia.

The effects of the LH surge on a suitably developed follicle are to bring about the final changes in the follicle which ultimately lead to ovulation and the formation of the corpus luteum.

Prior to ovulation there is a great increase in the follicular fluid volume. This causes the follicle to swell out from the side of the ovary. The mean follicular diameter just prior to ovulation is 23mm, so pre-ovulatory follicles can be easily seen without a microscope. The outer wall of the follicle becomes thinner and eventually ruptures. The oocyte still surrounded by the cumulus oophorus is slowly extruded. The fimbria collect the oocyte together with the cumulus mass and sweep them into the Fallopian tubes with the aid of cilia.

Just before ovulation the primary oocyte completes its first meiotic division. This division is unequal: half of the chromosomes and almost all the cytoplasm goes to one cell, the secondary oocyte. The other cell, the first polar body, consists virtually of just the remaining half of the chromosomes and is eventually discarded. The second meiotic division does not take place until fertilisation occurs.

Once the oocyte has been expelled at ovulation the remaining follicle collapses. The granulosa cells increase in size and form the steroid secreting cells of the corpus luteum. The cells of the corpus luteum synthesise progesterone and some oestrogen in response to stimulation by LH. They are no longer sensitive to oestrogen or FSH. The corpus luteum usually only exists for 10 to 14 days if pregnancy does not occur. After a set period it appears to be programmed to degenerate without any external influences.

The primordial follicle takes about 85 days to reach ovulation. Throughout the reproductive life of a woman, primordial follicles are continuously recruited to start the process of follicular development. However, only one follicle proceeds as far as ovulation each month. This means that no more than about 400 oocytes could possibly be ovulated, given that most women end their reproductive career between 45 and 55. This point is termed the menopause, and is signalled by menstruation ceasing to occur. At this time the ovary starts to atrophy, and the supply of oocytes is exhausted.

An ovarian cycle is the period between successive ovulations, and usually lasts between 24 and 32 days in humans. However the detection of ovulation is very difficult in the intact woman, so the cycles are measured from the first day of menstruation. Ovulation then typically occurs in the centre of the menstrual cycle, and the cycle can be conveniently divided into pre- and postovulatory phases. During the preovulatory period the follicles develop from the preantral state, and so this period is often referred to as the follicular phase. The postovulatory phase coincides with the existance of the corpus luteum, and so is alternatively called the luteal phase.

As already described the ovarian changes are accompanied by distinct patterns of hormonal secretion. The steroids, oestrogen and progesterone, secreted from the ovaries influence the release of LH and FSH from the anterior pituitary. This influence is mediated by GnRH. The precise role of the ovarian steroids is complex. In particular oestrogen elicits two different responses depending on its blood concentration. This can be explained if there are two separate "centres" within the hypothalamus governing GnRH release. One, the tonic centre, maintains relatively constant levels of LH and FSH. It is inhibited by oestrogens and hence oestrogen provides a negative feedback control. The second, the cyclic centre, is responsible for the midcycle LH surge. This is stimulated by oestrogens and so works on the basis of positive feedback. However this positive feedback system appears to be only

active when the oestrogens are present in sufficiently high concentrations, such as during the peak of oestrogen secretion from the preovulatory antral follicles.

Actions of ovarian hormones

OESTROGENS

- Develop and maintain female secondary sexual characteristics including the female shape
- Regulate gonadotrophin release
- Control follicular development
- Develop and maintain the uterus, Fallopian tubes, cervix, vagina, labia major and minor, and breasts
- Cause growth of the endometrium
- Produce changes in cervical mucus
- Increase motility of Fallopian tubes and excitability of uterine muscle
- Are involved in pregnancy

PROGESTERONE

- Produces secretory changes in endometrium (prepares uterus to receive embryo)
- Relaxes smooth muscle
- Creates secretory glands in breast
- Raises resting metabolic rate
- Raises basal temperature
- Inhibits release of gonadotrophins
- Maintains uterus during pregnancy

The two alternative ways that oestrogen influences the release of GnRH

Since release of ovarian hormones is cyclical, some of their effects occur with a periodicity related to the menstrual cycle. The most noticeable is the regular growth of the lining of the uterus (endometrium) followed by its degeneration when pregnancy has not started. Oestrogen causes endometrial growth, and progesterone brings about secretory changes. The endometrial glands become tortuous and dilated, and secrete a thick fluid rich in glycoprotein, sugars and amino acids. All this activity is preparing the uterine lining for implantation of a fertilised ovum. However if this does not occur, the endometrium breaks down and is shed together with blood from the torn arteries. This is menstruation and lasts 3-7 days in 95% of women. The blood lost during menstruation does not clot easily, and may vary from 20 to 200 ml for a single period.

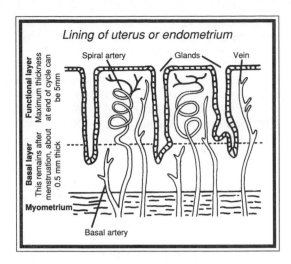

Lining of uterus or endometrium

Menstrual cycle

Fertilisation

Fertilisation usually takes place in the ampulla, the wide upper third of the Fallopian tube. The spermatozoa have to surmount a considerable number of barriers within the female reproductive tract to reach the ova in the ampulla. In so doing only a fraction of the enormous number of spermatozoa ejaculated into the vagina survive. Nevertheless the successful spermatozoa take just over an hour to reach the ampulla. They remain capable of fertilising for between 1 and 2 days. However, freshly ejaculated spermatozoa seem to be incapable of fertilising for the first 6 or 7 hours after release. A process known as capacitation takes place during this period. It involves a change in the surface coating of the spermatozoa, and requires the environment of the female reproductive tract.

During menstruation there is a rise in FSH secretion which causes follicle development. The maturing follicle secretes increasing amounts of oestrogen. This results in the midcycle surge of LH and FSH, and in turn leads to ovulation. Next the corpus luteum is formed and the presence of LH stimulates it to produce progesterone and some oestrogen. These corpus luteal steroids inhibit the release of LH and FSH. The corpus luteum then degenerates if fertilisation has not occurred. This removes the supply of progesterone and oestrogen, and thus evokes the degeneration of the endometrium and menstruation.

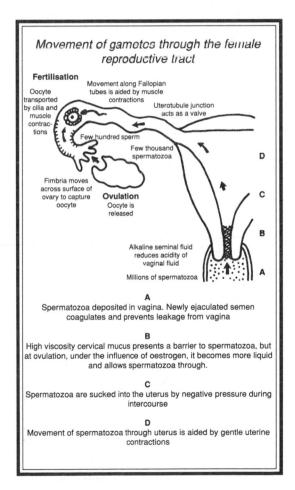

A
Spermatozoa deposited in vagina. Newly ejaculated semen coagulates and prevents leakage from vagina

B
High viscosity cervical mucus presents a barrier to spermatozoa, but at ovulation, under the influence of oestrogen, it becomes more liquid and allows spermatozoa through.

C
Spermatozoa are sucked into the uterus by negative pressure during intercourse

D
Movement of spermatozoa through uterus is aided by gentle uterine contractions

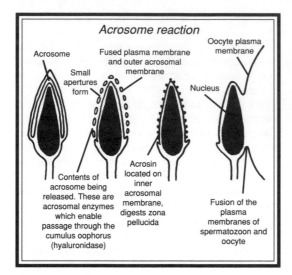

Following capacitation the acrosome reaction occurs. This releases enzymes from the acrosome which enables the spermatozoon to penetrate the cumulus cells and zona pellucida still surrounding the oocyte. The spermatozoon membrane then fuses with the oocyte plasma membrane.

Within 2-3 hours of the fusion, the oocyte completes its second meiotic division and releases the second polar body. The remaining 23 chromosomes of the maternal gamete can now combine with the 23 chromosomes of the paternal gamete to produce the zygote.

Initial development of embryo

Soon after the production of the zygote cleavage begins. This consists of repeated mitotic divisions of the zygote while it is still contained within the zona pellucida. During cleavage the total mass of cells does not change, and hence with each successive division the cell size gets smaller. Eventually a solid ball of 30-50 cells, the morula, is formed.

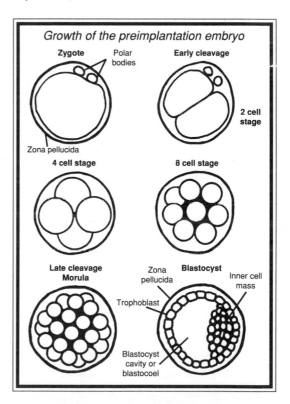

At about this time the morula has reached the uterine cavity after being transported along the Fallopian tube. The transport is aided by cilia and muscular contractions of the Fallopian tube.

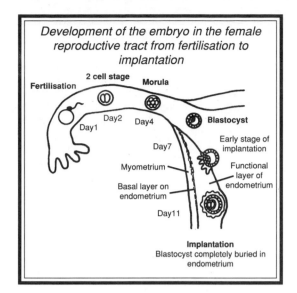

Development of the embryo in the female reproductive tract from fertilisation to implantation

will develop into extraembryonic membranes and part of the placenta. Hence it plays an essential role in the establishment and continuation of pregnancy. The inner layer of cells, the inner cell mass, forms the embryo proper.

Next the zona pellucida is shed, which until now has served to prevent the cells falling apart. If they do become divided into two distinct groups of cells, identical twins will result.

The blastocyst then adheres to the uterine epithelium. This usually takes place on the posterior wall of the uterus. The trophoblast secretes proteolytic enzymes which enable the blastocyst to sink into the uterine lining. Eventually the blastocyst is completely buried in the endometrium and implantation has occurred.

The trophoblast thickens and projects fingerlike processes into the surrounding endometrium. As this invasion proceeds, the maternal blood vessels become disrupted so that the trophoblast cells are in direct contact with maternal blood. This is the beginning of the formation of the placenta.

A fluid filled cavity then forms within the mass of cells, and the cells separate into inner and outer layers. This is the blastocyst. The outer layer of cells, the trophoblast, does not go on to form any part of the new individual. Instead the trophoblast

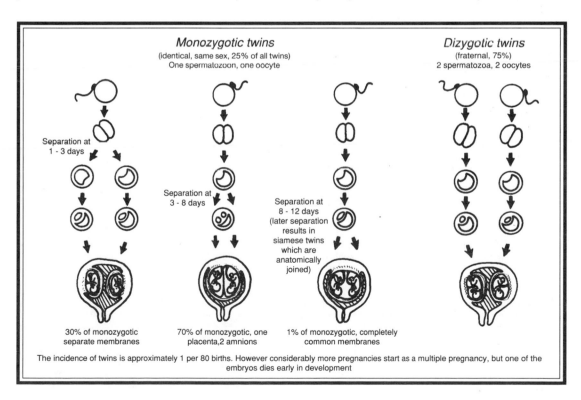

The incidence of twins is approximately 1 per 80 births. However considerably more pregnancies start as a multiple pregnancy, but one of the embryos dies early in development

Continuation of pregnancy

Pregnancy is now established with the implantation of the embryo. In order for it to continue, regression of the corpus luteum must be prevented. This is achieved by the secretion of the protein hormone, chorionic gonadotrophin (hCG). Initially prior to implantation the trophoblast secretes hCG, and continues for about 60 days. At this point the placenta is able to take over the role of the corpus luteum. The importance of the corpus luteum in maintaining the early pregnancy is its secretion of progesterone and oestrogen. A continued supply of progesterone is needed in order to prevent menstruation taking place. The amounts of progesterone and oestrogen in the blood rise throughout pregnancy. They inhibit the release of gonadotrophins and are essential for the advance of pregnancy. In particular they promote growth of the uterus and mammary glands.

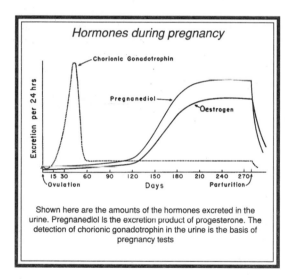

Hormones during pregnancy

Shown here are the amounts of the hormones excreted in the urine. Pregnanediol is the excretion product of progesterone. The detection of chorionic gonadotrophin in the urine is the basis of pregnancy tests

The embryo is derived from the inner cell mass of the blastocyst. Initially the inner cell mass forms a flat embryonic disc. By 14 days after fertilisation the cells have differentiated into three primary germ layers, an external layer or embryonic ectoderm; a middle layer or embryonic mesoderm; and an internal layer or embryonic endoderm. The cells of these layers continue multiplying in a set pattern to form layers, folds and bubbles of cells. Ultimately each of the germ layers forms into the various organs and tissues of the fetus.

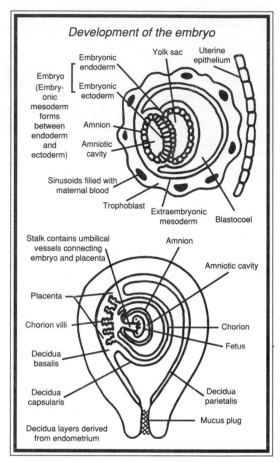

Development of the embryo

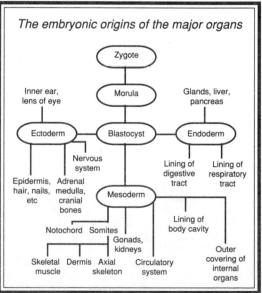

The embryonic origins of the major organs

The length of a normal human pregnancy is 40 weeks. This is measured from the first day of the last menstrual period, since most women can not date the act of coitus at which fertilisation happened. By six weeks into pregnancy the developing embryo measures 10 mm with a head and a tail end. The tube that will form the heart already shows pulsations. The embryo now changes rapidly and begins to look more and more like a baby. After eight weeks it is called a fetus, measures 25 mm and has limbs with toes and fingers. By twelve weeks all the organ systems have completed their primary development. The fetus has finger and toe nails, and measures 90 mm. From here on the changes that take place are mainly in growth.

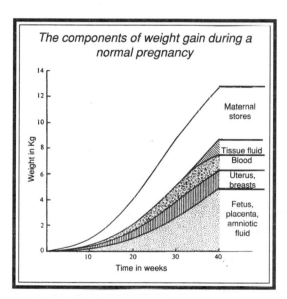

The components of weight gain during a normal pregnancy

Consequently by three months any congenital defects that are going to occur will already have happened. The first trimester is also the period during which the fetus is most sensitive to damage by drugs or viruses that cross the placenta.

While the embryo is developing, the trophoblast has differentiated into the fetal membranes around the embryo and the fetal component of the placenta. There are four of these fetal membranes, the yolk sac, allantois, amnion and chorion. The yolk sac is nonfunctional in humans. The allantois forms part of the placenta and the connection between the placenta and fetus, the umbilical cord. Meanwhile the amnion encloses the fetus floating it in the amniotic fluid. This fluid protects the fetus from dessication and mechanical shock. The fourth membrane is the chorion which surrounds the other three membranes and the fetus. Finally around the chorion is another layer derived from the maternal endometrium, the decidua capsularis.

Placenta

The placenta provides a means by which nutrients, gases and fetal waste products are exchanged between the mother and fetus. It is also an important source of hormones during pregnancy.

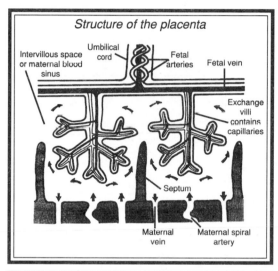

Structure of the placenta

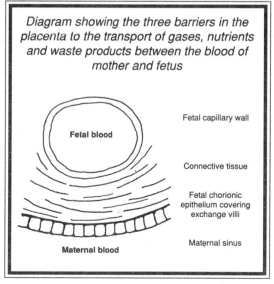

Diagram showing the three barriers in the placenta to the transport of gases, nutrients and waste products between the blood of mother and fetus

The placenta is a circular disc with a diameter of 15-20 cm and a thickness of 2-3 cm. It usually weighs about 500 gm. Within the placenta are branched villi of fetal tissue which dip into pools or sinusoids of maternal blood. There is no actual mixing of the fetal and maternal blood streams. In the human they are separated by three layers of fetal cells forming the placental barrier. The embryo is joined to the placenta by the umbilical cord containing two fetal arteries and a vein, which connect to capillaries in the placental villi.

Fetal circulation

Foramen ovale

RA
RV
Lungs (*)
LA
LV

Great veins

Ductus arteriosus

Aorta

Tissues

Ductus venosus

Liver

Digestive tract

Umbilical vein

Umbilical arteries

Placenta

Umbilical cord

Fetal bypasses which close at birth

(*) Only 10% of fetal blood passes through lungs

The placenta enables the fetus to use the maternal kidneys, skin, lungs and liver. Hence there are three shunts within the fetal circulation which ensure the optimal distribution of blood around the fetus. They close at or soon after birth. However closure of the foramen ovale, between the two atria of the heart, may take longer. It is still present in 10% of adults.

Only small molecules and gases can diffuse directly through the placental barrier. The transfer of oxygen is influenced by a number of factors including differential haemoglobin concentrations, patterns of placental blood flow and differences in the hae-moglobin molecule (see Chapter 3).

Larger molecules require transport mechanisms in order to pass through the placental barrier. The fetus needs various nutrients for growth. These include the essential amino acids, fatty acids, vitamins and minerals. However the major ingredient of the diet of a fetus is carbohydrate, mainly in the form of glucose. Near term the human placenta must transfer glucose to the fetus at a rate of more than 20 mg/min.

The placenta also serves as an immunological barrier. Only one type of antibody IgG is transported through to the fetus. These antibodies provide the baby with some immunological protection until its own immune system develops completely. This does not occur until sometime after birth. Further immunological protection is given by the presence of immunoglobulins, IgM and IgA, in breast milk. This supplies the immature gut with antibodies; breastfed babies have a better resistance to infections.

The placenta is an important source of hormones that regulate:-

• Rate of fetal growth.

• Muscular activity of the uterus. This prevents premature expulsion of the fetus and ensures that labour occurs at the correct time.

• Activity of the breasts.

The principal placental hormones are:-

• Chorionic gonadotrophin which prolongs the life of the corpus luteum.

• Oestrogens which stimulate uterine growth and development.

• Progesterone which quietens down the spontaneous activity of the uterus.

• Human placental lactogen or chorionic somatomammotrophin which is a protein hormone with a structure similar to growth hormone. It alters glucose and insulin metabolism.

Parturition

The actual process of expulsion of the fetus, its surrounding membranes and placenta from the uterus is called parturition. The precise mechanism which initiates parturition has not yet been clearly determined. It is likely to involve signals from the fetus itself, in the form of secretions such as cortisol from the fetal adrenal cortex. Other factors involved include prostaglandins, oestrogens, falling progesterone levels, oxytocin, the external environment and mechanical stretch of the uterus. The role of oxytocin is described in Chapter 13.

Once parturition has begun it will proceed as though controlled by a positive feedback mechanism, until the appropriate conclusion is reached.

At the beginning of parturition the cervix softens and dilates. Regular waves of muscle contraction pass over the uterus. These push the baby out through the uterus and vagina. The contractions occur once every 20 minutes to begin with, but as labour progresses they increase to once every minute. The cervix is then pulled up rather like a sock with a hole in the toe is pulled over a foot. The foot will go through the hole stretching the hole to accept its passage. The fetal head passes through the cervix in a similar manner.

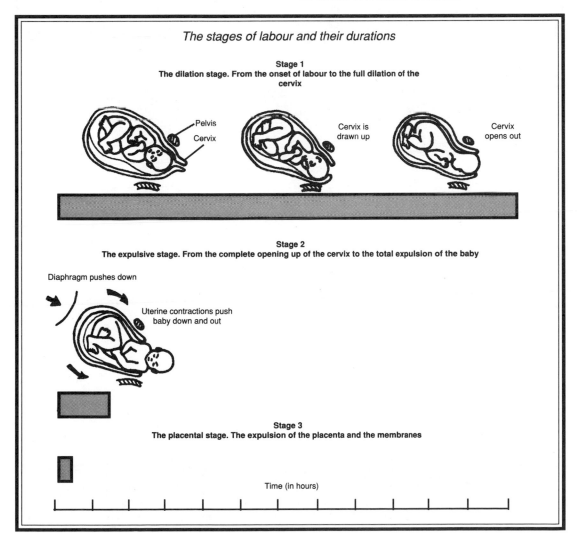

The stages of labour and their durations

Stage 1
The dilation stage. From the onset of labour to the full dilation of the cervix

Pelvis
Cervix

Cervix is drawn up

Cervix opens out

Stage 2
The expulsive stage. From the complete opening up of the cervix to the total expulsion of the baby

Diaphragm pushes down

Uterine contractions push baby down and out

Stage 3
The placental stage. The expulsion of the placenta and the membranes

Time (in hours)

Lactation

During pregnancy the breasts have been subjected to increasing levels of oestrogens and progesterone. The oestrogens increase the development of ducts, while the progesterone influences the proliferation of alveoli. By the fifth month of pregnancy the mammary gland is fully developed for lactation. However the initiation of full lactation does not occur until after parturition. The actual stimulus for the initiation is the removal of the high levels of oestrogens and progesterone from the maternal blood.

Once started the maintenance of lactation requires the hormone prolactin. Prolactin is released in response to suckling or nipple stimulation (Chapter 13). The actual ejection of milk relies on oxytocin (see also Chapter 13).

The composition of milk changes from an initial yellowish sticky secretion called colostrum, produced during the first week after birth, to true milk produced after 2 or 3 weeks. Colostrum contains more immunoglobulins and proteins, while mature milk has more lactose, more fat and a higher total calorific value.

Breast milk is the most suitable diet for young babies. Not only does it provide the correct nutrients, but it also supplies immunoglobulins. Breast feeding is also very satisfying to the mother, and helps establish the normal psychosocial interaction between mother and baby.

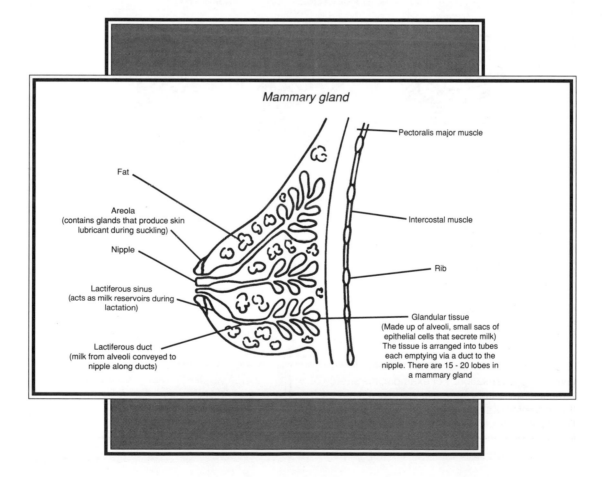

Mammary gland

Pectoralis major muscle

Fat

Areola
(contains glands that produce skin lubricant during suckling)

Nipple

Lactiferous sinus
(acts as milk reservoirs during lactation)

Lactiferous duct
(milk from alveoli conveyed to nipple along ducts)

Intercostal muscle

Rib

Glandular tissue
(Made up of alveoli, small sacs of epithelial cells that secrete milk)
The tissue is arranged into tubes each emptying via a duct to the nipple. There are 15 - 20 lobes in a mammary gland

RECEPTORS AND PAIN

SENSORY RECEPTORS

The central nervous system must be given complete information about any changes that occur in the environment.

General properties of receptors

Sensory receptors convert information regarding the environment into nerve impulses which are then transmitted to the central nervous system. There are many different kinds of sensory receptors; mostly they are specific for a given stimulus. There are receptors sensitive to light, heat, cold, sound, touch, pressure and stretch. In addition, there are receptors giving pain sensation.

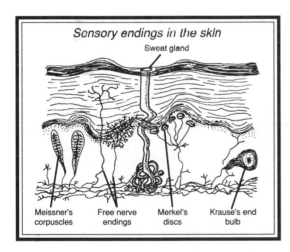

Sensory endings in the skin

Sweat gland

Meissner's corpuscles   Free nerve endings   Merkel's discs   Krause's end bulb

The stimulation of a receptor may not necessarily give rise to a sensation. For example, the stretch of a muscle activating a muscle spindle will not elicit the sensation of stretching, simply a reflex response takes place. Sensation is only generated when the impulses produced by a stimulated receptor travel to the sensory cortex.

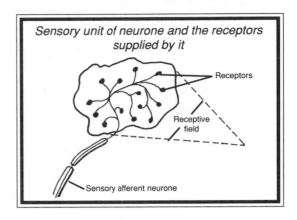

Sensory unit of neurone and the receptors supplied by it

Receptors

Receptive field

Sensory afferent neurone

Structure

The simplest receptors, such as those found in the skin, merely consist of a naked nerve terminal. Other more complex receptors have been found responding to such things as deep pressure or cold. Usually one sensory neurone supplies several receptors and this is known as a sensory unit which has a receptive field.

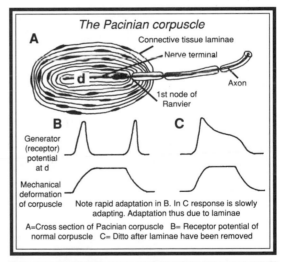

The Pacinian corpuscle

A

Connective tissue laminae
Nerve terminal

d

Axon

1st node of Ranvier

B   C

Generator (receptor) potential at d

Mechanical deformation of corpuscle

Note rapid adaptation in B. In C response is slowly adapting. Adaptation thus due to laminae

A=Cross section of Pacinian corpuscle   B= Receptor potential of normal corpuscle   C= Ditto after laminae have been removed

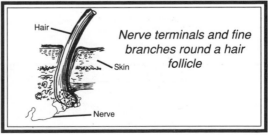

Hair

Skin

Nerve

Nerve terminals and fine branches round a hair follicle

Electrophysiology

Receptors are usually specific and have a very low threshold for a given type of stimulus. In general, receptors respond to their specific stimulus by producing a receptor (or generator) potential. This receptor potential is a non-propagated depolarisation in the terminal segment of the neurone.

If it is large enough a train of impulses is initiated in the nerve fibre. These are then transmitted to the central nervous system. The stronger the stimulus the larger the receptor potential and the higher the frequency of impulses.

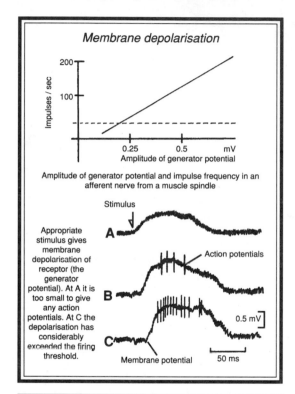

Membrane depolarisation

Amplitude of generator potential and impulse frequency in an afferent nerve from a muscle spindle

Appropriate stimulus gives membrane depolarisation of receptor (the generator potential). At A it is too small to give any action potentials. At C the depolarisation has considerably exceeded the firing threshold.

Adaptation

When a stimulus, which does not vary with time, is given to any receptor organ, the receptor potential that can be recorded from it, gradually decreases with time. This means that the frequency of impulses in the nerve fibre coming from the receptor also gets less. Of course, different receptors show this property of adaptation to different degrees. For example, touch receptors, hair receptors and particularly Pacinian corpuscles have rapidly adapting responses. Very often only one impulse is fired when a stimulus is given. On the other hand, some muscle and joint receptors have very slow adaptation. A given constant stimulus will produce a more or less constant firing rate in their afferent neurones. It is necessary for those endings which signal information about the position of limbs etc. to be of sufficient accuracy to represent the same joint angle every time that joint is bent. There are intermediate types between rapidly and slowly adapting endings; for example baroreceptors which signal blood pressure. The body needs to have information about blood pressure levels and a rapidly adapting organ would not signal the level at all; it would only signal the rate of change of blood pressure. These receptors showing intermediate properties have a dual response; a dynamic component which responds to the rate of change of stimulus, and a static component that signals the absolute level of the stimulus.

Adaptation of primary and secondary spindle endings

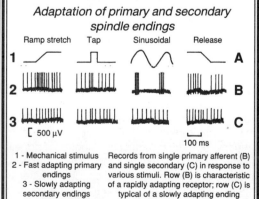

1 - Mechanical stimulus
2 - Fast adapting primary endings
3 - Slowly adapting secondary endings

Records from single primary afferent (B) and single secondary (C) in response to various stimuli. Row (B) is characteristic of a rapidly adapting receptor; row (C) is typical of a slowly adapting ending

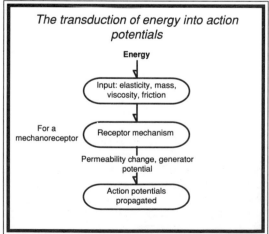

The transduction of energy into action potentials

For a mechanoreceptor

Energy

Input: elasticity, mass, viscosity, friction

Receptor mechanism

Permeability change, generator potential

Action potentials propagated

Just as in the case of motor information, the intensity of a stimulus is signalled in two ways:-

• Frequency of impulses, and the number of responsive receptors and therefore the number of active fibres

---

Types of sensory receptors

Mechanoreceptors
Skin touch, Deep pressure, Muscle stretch, Balance (labyrinthine receptors) spindles and Golgi organs, Sound, Blood pressure

Thermalreceptors
Skin temperature (hot and cold), Hypothalamic temperature

Chemoreceptors
Taste, Smell, Osmolality olfactory (hypothalamic receptors), Oxygen and Carbon dioxide, Blood glucose

EM Radiation
Light (retinal receptors)

Cutaneous receptors

The sensations that can be appreciated in the skin are touch, cold, heat and pain. Receptors can be demonstrated on skin areas by applying pressure with fine probes or using heated metal or water cooled devices. For all sensations there are small areas on the skin, unevenly distributed, that respond to different modalities.

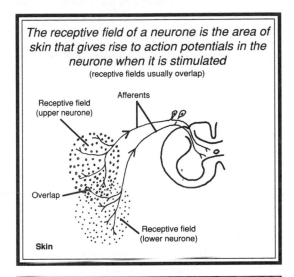

The receptive field of a neurone is the area of skin that gives rise to action potentials in the neurone when it is stimulated
(receptive fields usually overlap)

The particular structures that can be seen histologically on nerve endings in the skin do not correlate with particular sensory modalities. In the cornea there are no complex endings; only bare nerve fibres are present. However, the cornea has been found to give rise to sensations of heat, cold and pain. It may also be sensitive to touch, but it is difficult to distinguish between touch of the cornea and pain.

The specific nature of the message generated by cutaneous receptors is not so much a matter of the particular sensory ending that is stimulated, but it is the patterning of nerve fibre inputs that produces the sensation. It is not possible to stimulate skin electrically and produce, for example, sensations of heat or touch. Usually electrical stimulation elicits a peculiar sensation that most people describe as "pins and needles". This is the result of stimulating a whole group of nerve fibres together, which are not normally activated by natural means at the same time.

Two point discrimination

Skin receptors can be investigated by simultaneously stimulating two points on the skin. The distance between these points can be reduced to a minimum at which the stimuli feel separate. For example, if two sharp pin points are applied to skin of the middle of the back, they are felt as a single stimulus when they are 5 cm apart. However, the minimum separation on skin of the thumb or forefinger might be of the order of 1 or 2 mm. This gives a rough idea of the sensitivity of various skin regions to the particular stimulus involved.

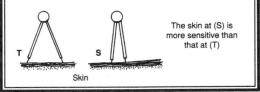

The skin at (S) is more sensitive than that at (T)

Temperature receptors

There are two kinds of temperature receptor. The cold receptors show maximum sensitivity to temperatures of 25-30°C (below deep body temperature). Warm receptors respond to temperatures higher than normal body temperature.

The ranges over which these hot and cold receptors work do overlap a bit. Moreover, it is possible to find a single receptor (or at any event the nerve fibre coming from it) that shows a bimodal response. In other words, the same receptor responds to both hot and cold.

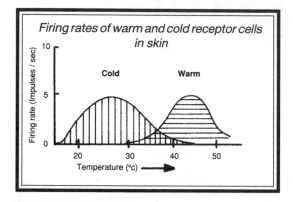

Firing rates of warm and cold receptor cells in skin

Receptors in skin sensation

Receptor	Modality	Fibre Dia (μm)
Hair follicle	Light touch	Aβ (5 – 12)
Meissner's corpuscle	Light touch	Aβ (5 –12)
Ruffini end organ	Skin steady pressure	Aβ (5 –12)
Free nerve ending	Pain and temperature	Aδ; C (<1 - 5)

Proprioceptors

Proprioceptors signal body position. They are in areas deeper than skin, and include joint receptors (signalling the degree of flexion of a joint) and muscle receptors. There are also receptors in tendons and ligaments.

Visceral receptors

The various internal organs are richly supplied with receptor endings and afferents. These receptors mainly respond to pressure and pain; some are chemoreceptors. Many of them give rise to reflex responses but their representation of the normal sensory modalities is rudimentary.

Some receptors are associated with the circulatory system and provide a measure of vascular pressure. Stretch receptors also occur in hollow viscera such as the bladder, intestinal tract and lungs.

Chemoreceptors are sensitive to chemical changes; most of them are in direct contact with circulating blood and measure such parameters as partial pressure of oxygen or acidity.

PAIN

Everyone is familiar with pain, yet research has yet to come up with a satisfactory definition. Pleasure and pain are driving forces in behaviour, but very little is known about them. The original view was that pleasure resulted from stimulation of "beneceptors" and pain from "nociceptors".

Psychology of pain

Some people, congenitally, do not feel pain. These persons have extensive burns, injuries of all kinds and often bite their own tongues. They usually have severe lesions in the spine, bones and joints, and die at an early age from infection. There are other people who have exaggerated pain sensibility. Severe pain may follow injury or amputation and is known as a phantom limb. Spontaneous pain may also occur.

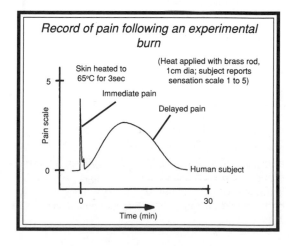

Record of pain following an experimental burn

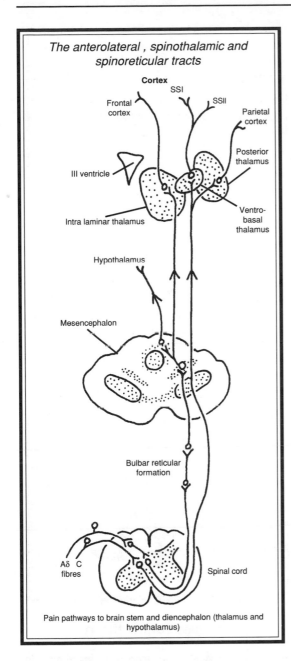

The anterolateral , spinothalamic and spinoreticular tracts

Cortex

SSI

Frontal cortex

SSII

Parietal cortex

Posterior thalamus

III ventricle

Ventro-basal thalamus

Intra laminar thalamus

Hypothalamus

Mesencephalon

Bulbar reticular formation

Aδ C fibres

Spinal cord

Pain pathways to brain stem and diencephalon (thalamus and hypothalamus)

Physiology of pain

It would appear to be a simple thing to investigate the mechanism of pain, by giving a painful stimulus and following the impulses generated through nervous pathways. This turns out not to be the case. Moreover it is impossible to stimulate pain receptors on their own. It is unknown whether the signalling of pain involves impulses in a specific pain pathway or whether the brain makes use of a whole pattern of afferent activation which is produced by noxious stimulation.

Pain might be due to over stimulation of any kind of sensory ending. Alternatively there might be specific pain endings. It would appear that the truth is somewhere between these two.

Pain receptors

There are four classes of pain receptor:-

- High threshold mechanoreceptors (myelinated A fibres)

- High threshold mechanoreceptors (non-myelinated C fibres)

- Polymodal nociceptors i.e. they signal more than one modality (non-myelinated C fibres). These are the most common.

- Heat and mechanically sensitive cells (thinly myelinated)

Synaptic connections of nociceptors

There are two types of cell in the spinal cord activated by painful skin stimulation: specific cells and cells which respond just as well to other stimuli (wide dynamic or common carrier units).

Higher centres

Pain fibres run in the spinoreticular and spinothalamic pathways. These have a widespread origin in cells throughout the spinal cord.

It has been generally held that pain fibres end in a "pain centre" in the brain. However, this cannot

Most people have a very similar threshold to sensation; an electric shock, that will give a just detectable sensation, is about the same voltage for everyone. However, the situation outside a laboratory is much more complicated. For example, during a battle a soldier may get severe injury without being aware of it.

explain a good deal of the contradictory findings about pain sensation (unless of course one calls the whole brain a pain centre). The thalamus, hypothalamus, brain stem, limbic system and cerebral cortex are all involved in pain perception. A tumor of the thalamus may give rise to severe, continuous, intractable pain (thalamic syndrome). On the other hand, a lesion of the frontal cortex may abolish the unpleasant aspects of severe pain and a patient with such a lesion very rarely complains about pain.

Theories for pain

Specific theory

The "specific theory" proposes that there are specific pain receptors in tissues which send impulses along pain fibres to a pain centre in the brain. These receptors are supposed to be free nerve endings connected to A and C nerve fibres which ascend in the anterolateral spinothalamic tract to the thalamus. This theory is untenable for the following reasons:-

• C fibres and their endings are stimulated by many kinds of innocuous stimuli as well as by painful ones.

• There are very few sensory fibres solely activated by noxious stimuli.

• Operations have been carried out to relieve intractable pain with the aim of cutting pain pathways. The most successful is cordotomy; here the whole of the anterolateral columns are divided. Loss of pain sensitivity on the opposite side of the body is produced, but usually this loss is very short lived (1 or 2 months). The conclusion is that pain pathways are variable.

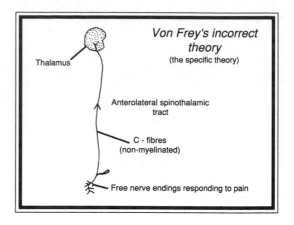

Von Frey's incorrect theory
(the specific theory)

Thalamus

Anterolateral spinothalamic tract

C - fibres (non-myelinated)

Free nerve endings responding to pain

The pattern theory

Pain sensation is induced by specific patterns of ascending nerve impulses generated by intense stimulation of any kind of receptor. In terms of pain, this implies that nerve endings and nerve fibres are non-specific. However, there is very good evidence that specialised endings and fibres do exist for signalling various different sensory modalities, including pain.

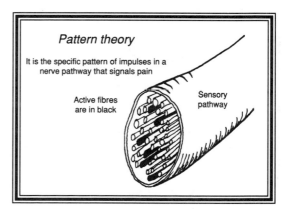

Pattern theory

It is the specific pattern of impulses in a nerve pathway that signals pain

Active fibres are in black

Sensory pathway

The gate control theory

Melzack and Wall introduced the gate theory in 1965, whereupon it was immediately hailed as a breakthrough. However, a number of discrepancies became obvious and it is now not believed. Nevertheless , a considerable modification of pain sensation does occur at the spinal level. It appears likely, that the main part of this modification originates as down coming nerve activity in the spinal

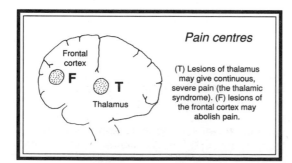

Pain centres

Frontal cortex

F   T

Thalamus

(T) Lesions of thalamus may give continuous, severe pain (the thalamic syndrome). (F) lesions of the frontal cortex may abolish pain.

cord. This controls the input from spinal to cortical and thalamic levels, like any other sensory modality.

Diagram of gate control theory (1965)

S.G. = Substantia gelatinosa cell of spinal cord

L = large fibres   S = small fibres

The inhibitory (gate) effect of S.G. cells upon T cells (first central transmission cells) in spinal cord is increased by activity in L fibres and decreased by S fibres. Central control is inhibitory feedback via descending pathways to the whole of the gating system. If the T cell output gets large enough, then pain is felt.

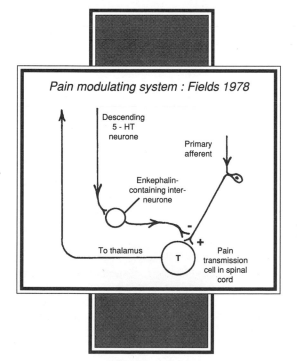

Pain modulating system : Fields 1978

Localisation of pain

The accurate location of a source of pain is biologically important, if this could lead to positive action for survival. Therefore accurate localisation is usual when pain arises in a superficial structure, such as the skin, where something can be done by the sufferer (e.g. removing a thorn). However, localisation is less accurate from bones and joints, and pain from viscera is usually not localised at all. Precise localisation and subjective analysis of pain is a function of the cerebral cortex.

Referred pain

Pain is often felt in an area remote from that which is being stimulated; usually this area is supplied by the same, or an adjacent, segment of the spinal cord. Patients with heart disease frequently have pain in the shoulder and root of the neck, although it originates in heart muscle. Often there are patterns of "trigger" areas in the chest and shoulder. Pressure on these may give an intense pain lasting for hours. Normal people have similar trigger areas, which also give discomfort when pressed but not severe or long-lasting pain. These areas are consistent from person to person and enable accurate diagnosis of disease. The injection of local anaesthetic into a trigger area abolishes referred pain.

Another example of referred pain is the pain of appendicitis. This occurs in the mid-line to begin with and not in the right iliac fossa as might be expected. Commonly, the diagnosis is missed initially and it is not until the appendix is about to burst, and the parietal layer of peritoneum is inflamed, that pain is felt over the site of the appendix.

SPECIAL SENSES

VISION

Light is electromagnetic energy which travels at 3×10^8 m/s. The eye detects light of wavelengths between 700 nm (red) and about 350 nm (violet).

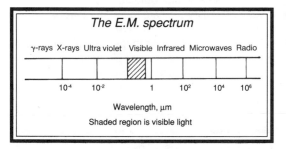

The E.M. spectrum

γ-rays X-rays Ultra violet Visible Infrared Microwaves Radio

Wavelength, μm

Shaded region is visible light

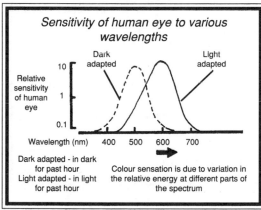

Sensitivity of human eye to various wavelengths

Relative sensitivity of human eye

Dark adapted

Light adapted

Wavelength (nm)

Dark adapted - in dark for past hour
Light adapted - in light for past hour

Colour sensation is due to variation in the relative energy at different parts of the spectrum

Formation of retinal image

The eye is constructed like a camera, the optical system of the cornea and lens being the equivalent of the camera's lens, the iris (and pupil) having the same function in both. The retina has the image formed upon it.

The refractive index (RI) of aqueous and vitreous humour is about 1.35, and that of the lens is about 1.43. It follows, therefore, that most of the refractive

Dark and light adaptation

The eye works well at an extraordinary range of brightness values (a dynamic range of 10^{15} units) due to its power of adaptation. This process of adaptation takes a certain amount of time to effect; it may take 30 minutes to adjust the eye to low levels of illumination after being in bright light. This adaptation can be measured as a person's absolute threshold to light.

A subject is exposed to strong light until time 0. The brightness of the dimmest light he can just perceive is recorded every 0.5 min, and plotted against time as an adaptation curve. This has two distinct components, the first part flattens out after 9 min and is due to cones. The second part lasts to about 30 min, and is due to the recovery of the rods which are sensitive to much lower illumination levels.

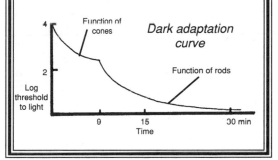

Function of cones

Dark adaptation curve

Function of rods

Log threshold to light

Time

power of the eye lies in the outside (anterior) surface of the cornea (relative RI = 1.35). The lens provides fine adjustment in power (i.e. focussing: relative RI = 0.08).

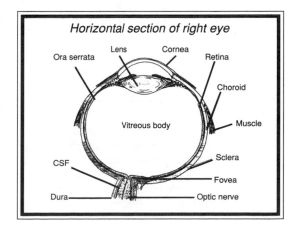

Horizontal section of right eye

Ora serrata
Lens
Cornea
Retina
Choroid
Muscle
Vitreous body
Sclera
CSF
Fovea
Dura
Optic nerve

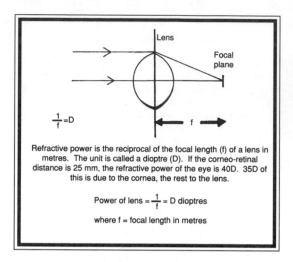

Refractive power is the reciprocal of the focal length (f) of a lens in metres. The unit is called a dioptre (D). If the corneo-retinal distance is 25 mm, the refractive power of the eye is 40D. 35D of this is due to the cornea, the rest to the lens.

$$\text{Power of lens} = \frac{1}{f} = D \text{ dioptres}$$

where f = focal length in metres

Accommodation

Although only about 5D of the total power of the eye is due to the lens during distant vision, the lens can alter its shape and so increase its power. This focusses images on the retina, a process called accommodation. An additional 10-15D can be produced. The lens is held radially by a suspensory ligament attached to the ciliary muscle arranged in an annulus around the corneo-scleral junction. When the ciliary muscle relaxes, a radial pull is exerted on the lens which becomes flatter.

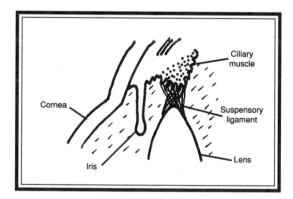

Focussing on a near object (accommodation) is brought about by contraction of the ciliary muscle; the natural elasticity of the lens forces it to become more globular. Presbyopia is a condition in old people whereby this elasticity is lost and the focussing power may be reduced to 1D or less.

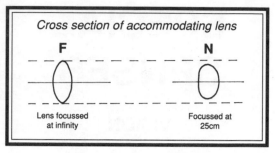

Cross section of accommodating lens

F — Lens focussed at infinity

N — Focussed at 25cm

Visual defects

When the lens brings objects into sharp focus on the retina, the eye is said to be emmetropic. Less than 5% of the population have normal vision, most being either long-sighted (hypermetropic) or short-sighted (myopic). Some people have, in addition, astigmatism, a condition in which the lens is slightly cylindrical, bringing vertical or horizontal lines to a focus in differing planes.

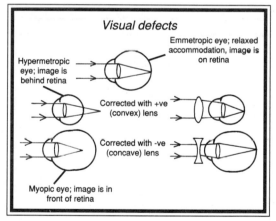

Visual defects

Pupil

The iris diaphragm is controlled by:-

(a) The sphincter pupillae, a circumferential muscle whose contraction constricts the pupil.

(b) The dilator pupillae, a radial muscle whose contraction dilates the pupil.

Muscles (a) and (b) have opposing actions; they are controlled by the parasympathetic and sympathetic systems respectively. The pupil is dilated by atropine which blocks acetylcholine. It is also dilated by adrenalin or adrenergic drugs such as phenylephrine(see Chapter 16).

As in the camera, constricting the pupil increases depth of focus. Most people with minor visual defects can thus see more clearly in bright light when the pupil is constricted. The pupil accounts for only 1.2 log units of the adaptation to intensity of illumination.

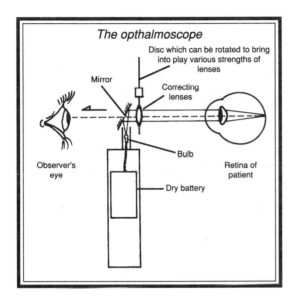

The opthalmoscope

Retina

The retina can be directly examined through the lens using an opthalmoscope. This consists essentially of a series of correcting lenses (to correct for errors of refraction in the subject and observer) and a small source of illumination.

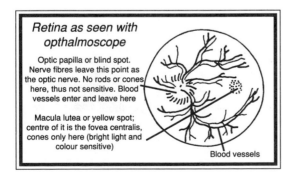

Retina as seen with opthalmoscope

Optic papilla or blind spot. Nerve fibres leave this point as the optic nerve. No rods or cones here, thus not sensitive. Blood vessels enter and leave here

Macula lutea or yellow spot; centre of it is the fovea centralis, cones only here (bright light and colour sensitive)

Blood vessels

Blind spot

Nerves and blood vessels leave the retina at the blind spot. This area does not respond to light as there are no receptors. It is 5 deg across, yet one is unaware of it.

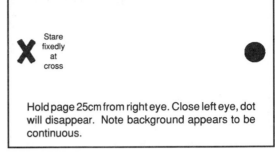

Stare fixedly at cross

Hold page 25cm from right eye. Close left eye, dot will disappear. Note background appears to be continuous.

Fovea centralis

The fovea centralis is the region of maximum density of cones (no rods). When an object is looked at, the eyes are directed so that its image falls on the fovea.

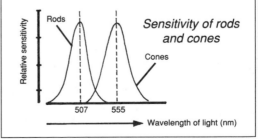

Rods and cones

There are two kinds of photosensitive cell in the retina. Cones are responsible for vision in bright light. They work in daytime vision and give rise to detailed images in colour.

Rods are responsible for vision in dim conditions. They respond to very small amounts of light, but do not subserve colour or sharp images. The retina is responsive to wavelengths between 350 and 700 nm. However, the sensitivity is not equal at all wavelengths (colours). Cones respond best to yellow-green light (peak at 555nm); rods to green-blue light at 507nm.

Sensitivity of rods and cones

Visual pigment

Rhodopsin is responsible for the absorption of light. It is composed of a pigment, retinal, which has a purple colour when it combines with the protein, opsin. If light falls on this pigment, the light is absorbed and the retinal is released from the protein. This changes the membrane potential of the rod, which eventually leads to the production of action potentials in the optic nerve. The rod receptor potential, unlike receptor potentials generated elsewhere, is a hyperpolarisation. When the stimulus is over, retinal again combines with opsin, and rhodopsin is regenerated. Some of the retinal, however, becomes converted into vitamin A and diffuses into the blood stream, where it is removed. If blood levels of vitamin A are low, then the replacement of retinal is incomplete and the action of rods is impaired. This condition is night blindness.

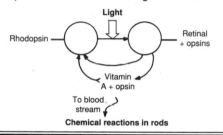

Chemical reactions in rods

Rhodopsin in daylight

Under conditions of strong illumination, most of the rhodopsin remains broken down and rods are inactive. In these circumstances a person going from bright to dim illumination is at first blind. This is because little rhodopsin is present. As time passes, however, in dim light rhodopsin becomes regenerated and vision improves (dark adaptation).

Colour vision

Rods are not sensitive to colour. Hence at night, colours of objects are not perceived. Colour vision is the function of cones, which are less sensitive than rods. About 200 colours can be made out by the normal eye. This is achieved by only 3 types of visual pigment in the cones. These visual pigments are chemically related to rhodopsin and they work in exactly the same way.

Although all cones appear identical, each one is selectively sensitive to only one of the three primary colours, red, green and blue. The colour perceived by the cerebral cortex depends on the proportion of the three types of cones being stimulated. For example, light reflected from grass is very largely composed of wavelengths that will maximally activate the green sensitive cones. For other colours either single types of cone or a mixture will be stimulated. Whiteness, as a sensation, occurs when the three kinds of cones are stimulated to roughly equal extents.

Visual nervous pathways

There are around 150 million rods and cones in the retina and information from these is transmitted to the visual cortex. The primary receptors connect with bipolar cells and they, in turn, activate ganglion cells. Ganglion cell axons leave the retina and become the optic nerve which contains about one million nerve fibres. The information is then passed onto the cortical visual areas (occipital cortex) via the lateral geniculate nucleus where synaptic contact is made with spinal neurones ending in the cortex.

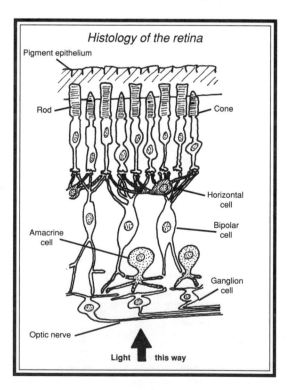

Histology of the retina

Several rods and cones converge on a single bipolar cell and several bipolar cells activate one ganglion cell. Thus, the retinal field from which one bipolar cell is activated is much larger than a single receptor cell. The retinal field of a ganglion cell is larger still. Retinal fields of all these cells overlap.

At each stage in the visual pathway information is processed and the cells at each successive relay station respond to ever more complex patterns of light falling upon their receptive fields. Ganglion and lateral geniculate cells have circular receptive fields and respond to spots of light falling within them. Whereas the "simple" cells of the visual cortex respond to light or dark, straight bars or edges of specific orientations. The "complex" cells respond to similar stimuli which are moving, while the "hypercomplex" cells are stimulated by more complicated shapes with specific orientations moving across their receptive fields.

Visual acuity

The term visual acuity means the eye's ability to perceive clearly. In other words, it is the resolving power of the eye, measured by the distance at which the eye can separate two lines drawn close together. In standard visual acuity tests the patient is asked to read letters placed at a distance 20 feet away from him.

HEARING

Sound

Sound is a vibration of air molecules, alternate

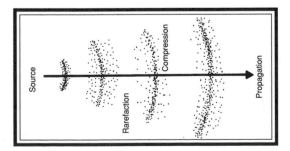

> Frequency, F=cycles/second (or Hertz, Hz)

compressions and rarefactions being propagated at about 330 m/s.

Sound quality

This is the sensation generated by a mixture of frequencies.

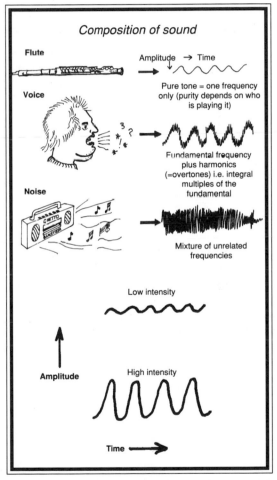

Composition of sound

Intensity
The strength of perception depends on the degree of fluctuation in the pressure.

Audibility

This is the psychological perception of sound. It is proportional to the logarithm of intensity. The ear is most sensitive at between 2 and 3 kHz.

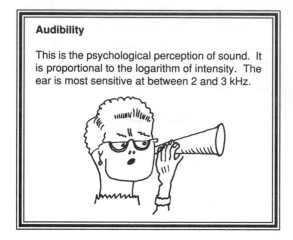

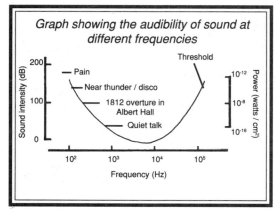

Graph showing the audibility of sound at different frequencies

Decibels

To measure this very large range of intensity, the unit of intensity, or decibel, was adopted. It is 10 x the log of the ratio of the intensity in question to a reference sound.

Physiology of hearing mechanisms

The hearing sensory elements are in the Organ of Corti and are stimulated by movements of the basilar membrane. Sound in air has to be transformed into sound in water which is a very much denser medium. The eardrum is vibrated by a sound in the environment. This vibration is

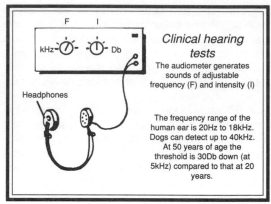

Clinical hearing tests

The audiometer generates sounds of adjustable frequency (F) and intensity (I)

The frequency range of the human ear is 20Hz to 18kHz. Dogs can detect up to 40kHz. At 50 years of age the threshold is 30Db down (at 5kHz) compared to that at 20 years.

transformed by a series of levers (malleus, incus, stapes) to a much smaller amplitude movement at the oval window.

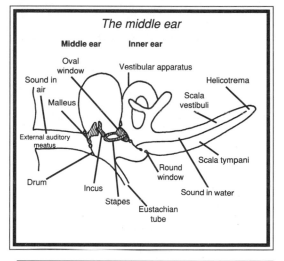

The middle ear

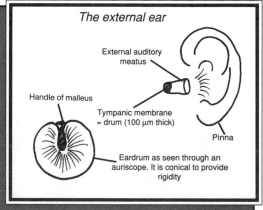

The external ear

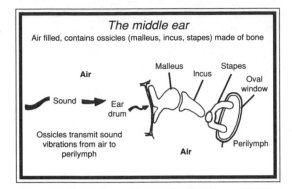

The middle ear
Air filled, contains ossicles (malleus, incus, stapes) made of bone

Air

Sound ➡ Ear drum

Malleus
Incus
Stapes
Oval window
Perilymph
Air

Ossicles transmit sound vibrations from air to perilymph

Impedance matching

The function of the middle ear is to couple movements of low density air to the high density perilymph, in which the basilar membrane is bathed. Thus high amplitude - low force air vibration is transformed into a low amplitude - high force liquid vibration. This is achieved by the area of the drum being 0.7 cm^2 while the oval window is 0.02 cm^2.

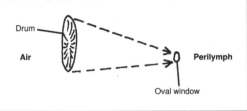

Drum

Air

Perilymph

Oval window

Protective actions

The middle ear also acts to protect the inner ear from very loud sounds.

(1) Rotation of stapes occurs if its movement is too large, whereas at low sound levels it hinges.

(2) Reflex action of tensor tympani muscle. A loud sound reflexly pulls the drum towards the inner ear.

(3) Reflex action of stapedius muscle. This muscle pulls the stapes out of the oval window.

Mechanisms (2) and (3) are slow, while (1) is instantaneous. They enable the ear to utilise an enormous dynamic range, pressures being up to 10^6 times threshold.

Protective responses to loud sounds

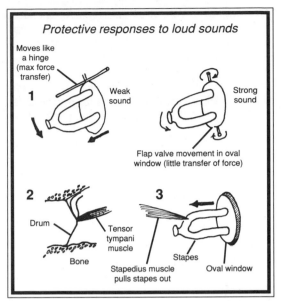

Moves like a hinge (max force transfer)

1   Weak sound   Strong sound

Flap valve movement in oval window (little transfer of force)

2   3

Drum

Tensor tympani muscle

Bone

Stapedius muscle pulls stapes out

Stapes

Oval window

Eustachian tubes

These connect the middle ear with the pharynx so that pressure does not build up in the middle ear. Airline passengers have problems if the tubes are blocked with catarrh. Yawning and swallowing opens them.

Inner ear

Movement of the stapes causes pressure changes in the perilymph. The basilar membrane thus moves; in turn this gives rise to impulses in the VIIIth nerve.

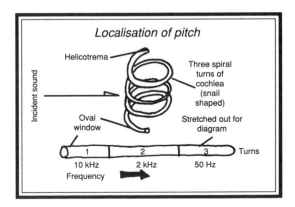

Localisation of pitch

Helicotrema

Incident sound

Oval window

Three spiral turns of cochlea (snail shaped)

Stretched out for diagram

1   2   3   Turns
10 kHz   2 kHz   50 Hz
Frequency

The diagram shows where pitch discrimination is localised along the cochlea. Measurements were made by von Bekesy (1960) with silver particles applied to the basilar membrane and their movement observed under a microscope when sounds of various frequencies were switched on. Low frequencies are localised at the far end near the helicotrema; high frequencies at the near end.

Resonance theory (Helmholtz)

This theory pictured the basilar membrane acting like strings in a harp. However, the large fluid mass and the known low tensions in the basilar membrane do not fit this theory, so it is inadequate.

At low frequencies the whole membrane moves and frequency is discriminated by the auditory cortex. Above 100 Hz, only part of the basilar membrane is set in motion by sound; this is "place" discrimination. Above 10 kHz discrimination is very poor anyway.

Initiation of action potentials in VIIIth nerve

A 3 kHz, just threshold, sound moves the tympanic membrane 10^{-9}cm (1/10 diameter of a hydrogen atom!). The movement of the basilar membrane is even less than this (Why don't we hear the blood flowing in the tympanic membrane?)

Relative movement sets up shear stress in the hairs, but the mechanism of generation of nerve impulses is unknown.

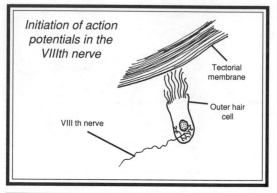

Initiation of action potentials in the VIIIth nerve

Tectorial membrane

Outer hair cell

VIII th nerve

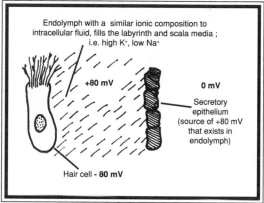

Endolymph with a similar ionic composition to intracellular fluid, fills the labyrinth and scala media ; i.e. high K^+, low Na^+

+80 mV

0 mV

Secretory epithelium (source of +80 mV that exists in endolymph)

Hair cell - 80 mV

Receptor potential

There is about 160 mV potential difference across the hair cell membrane. Probably the hair cells are more sensitive to deformation for this reason, and the curving of the hairs might lead to the "opening of pores".

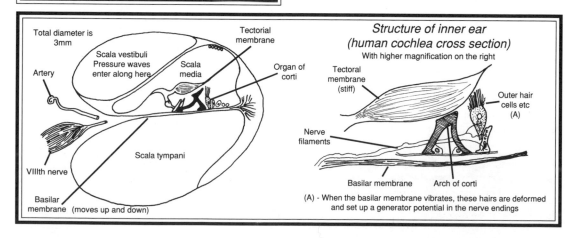

Total diameter is 3mm

Scala vestibuli Pressure waves enter along here

Artery

Scala media

Tectorial membrane

Organ of corti

Scala tympani

VIIIth nerve

Basilar membrane (moves up and down)

Structure of inner ear (human cochlea cross section)
With higher magnification on the right

Tectoral membrane (stiff)

Outer hair cells etc (A)

Nerve filaments

Basilar membrane   Arch of corti

(A) - When the basilar membrane vibrates, these hairs are deformed and set up a generator potential in the nerve endings

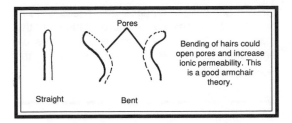

Bending of hairs could open pores and increase ionic permeability. This is a good armchair theory.

Straight　　　Bent

Sound intensity is signalled by frequency of impulses in the VIIIth nerve; the higher the intensity, the higher the impulse frequency.

Localisation of sound source

There are several mechanisms to localise a sound source:-

(1) Time of arrival of pressure point in each ear. Most people can detect a time difference of 0.05-0.1 ms.

(2) Localisation depends on the phase difference in the two ears. This is a similar mechanism to (1), but works with continuous sound and is best with pure sound waves. Humans can best localise continuous sounds at around 2 kHz. Below this the wave length gets too long; above it is too short.

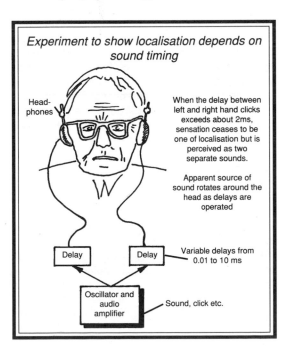

Experiment to show localisation depends on sound timing

Head-phones

When the delay between left and right hand clicks exceeds about 2ms, sensation ceases to be one of localisation but is perceived as two separate sounds.

Apparent source of sound rotates around the head as delays are operated

Delay　　　Delay

Variable delays from 0.01 to 10 ms

Oscillator and audio amplifier

Sound, click etc.

(3) Intensity in each ear aids localisation. The ear away from the sound receives a lower intensity. Experiments show that differences of about 1 dB can be distinguished. The shadowing effect may be up to 25 dB depending upon frequency, waveform, etc. This mechanism works best at high frequencies, because then the shadow, cast by the head, is sharper.

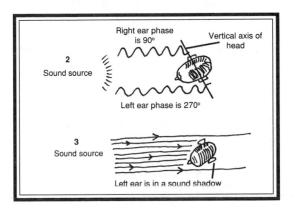

Right ear phase is 90°

Vertical axis of head

2 Sound source

Left ear phase is 270°

3 Sound source

Left ear is in a sound shadow

VESTIBULAR SYSTEM

The vestibular system is a receptor system for balance, being the non-acoustic part of the inner ear. The semi-circular canals monitor angular head movement; the utricle and saccule (otolith organs) detect position and linear motion.

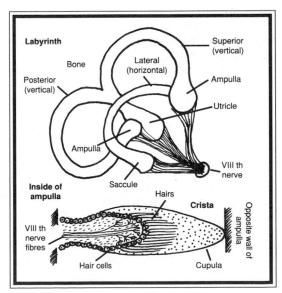

Labyrinth

Superior (vertical)

Bone

Lateral (horizontal)

Posterior (vertical)

Ampulla

Ampulla

Utricle

VIII th nerve

Saccule

Inside of ampulla

Hairs

Crista

Opposite wall of ampulla

VIII th nerve fibres

Hair cells

Cupula

Structure of labyrinth

The semi-circular canals are continuous with the utricle and are filled with endolymph. Each canal has a sensory region, the ampulla. The receptor organ here is the crista which is a ridge of columnar epithelial cells, whose hairs are embedded in a gelatinous mass, the cupula. The ridge touches the opposite wall of the ampulla and will be moved if there is any flow of endolymph.

Stimulation of semi-circular canals

Under conditions of no movement, the basal impulse frequency in the vestibular nerve filaments is the same on each side of the head (originating in all the ampullae).

• Horizontal canals: if the cupula is bent towards the utricle, nerve impulse frequency is increased (positive stimulation); when bent away, it is decreased.

• Vertical canals: these are the opposite to the horizontal canals. Thus any motion of the cupula from resting will produce a discharge, the prime effect of which is a sensation of rotation.

TASTE AND SMELL

Taste and smell are similar modalities; food has a flavour which is usually the result of stimulation of both kinds of receptors. This is clear when a cold in the head blocks nasal passages, and food loses much of its taste. In humans chemical senses are relatively undeveloped and smell is much less important than it is in some other animals.

Taste

There are four primary taste sensations: sweet, sour, salt and bitter. Chemical substances giving rise to the sensation of sourness are acids and the degree of sourness depends on pH. Substances tasting salty are chemical salts such as sodium chloride. Sweetness arises from stimulation by many chemicals; organic molecules are sweet e.g. sugars, amino acids. Bitter substances include

quinine and salts of magnesium. These have no clearly related chemical structure. It turns out that many poisons are bitter and taste receptors are particularly sensitive to bitterness; presumably this is of a protective nature.

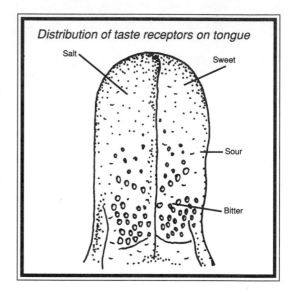

Distribution of taste receptors on tongue

Salt — Sweet — Sour — Bitter

Taste buds

Taste receptors are small ovoid structures, most of which are on the tongue, but they are also found in the pharynx and on the palate. Each of these taste buds has 20 to 30 receptor cells with thin filamentous tips projecting through a pore. The filaments act as the receptor surface, chemical stimulation of which gives rise to a receptor potential and eventually to propagated impulses in afferent nerve fibres. One receptor cell does not respond specifically to one kind of taste, although in most cases it will be more sensitive to one of the four tastes. Taste buds at the tip of the tongue are most sensitive to sweet substances; those at the back and side to the other taste modalities.

Fibres from taste buds travel in cranial nerves to the brain stem and impulses are then passed to the thalamus and onto the cortex. The cortical area involved is near to the tongue region in the somatotopic map. There are connections which also run to salivation centres in the medulla; the autonomic innervation of salivary glands is reflexly activated via this pathway.

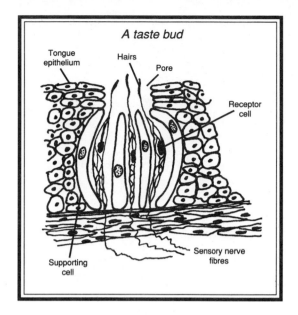

A taste bud

Tongue epithelium

Hairs

Pore

Receptor cell

Supporting cell

Sensory nerve fibres

Olfaction shows rapid adaptation. Pleasant and unpleasant smells rapidly disappear with time, even though the odour itself remains present. This adaptation is specific for a given odour. There is no change in the threshold for other odours which may appear. Such adaptation takes place mainly in the central nervous system, but also involves the receptors.

Smell

In the nasal cavity lies the olfactory membrane. This is above the air stream passing in and out of the nose, but it is reached by diffusion and eddy currents. These eddies are set up by sniffing. Millions of receptors are present in the olfactory membrane. They are modified ciliated cells, their cilia projecting into the mucus that lies on the nasal mucosa. Axons from these receptor cells form the olfactory nerve. Only volatile substances can be smelled because unless a chemical can vaporise to an extent great enough to be present in the air breathed in, it cannot produce an olfactory stimulus. Moreover, it must be sufficiently soluble in water to dissolve in the mucus and hence stimulate the cilia.

Large numbers of different odours can be appreciated; probably up to about 10,000. Therefore the olfactory sense is a good deal more complex than that of taste.

The sensory pathway for olfaction goes to the olfactory bulb and thence to the olfactory cortex beneath the cerebral cortex. It is the only sensory pathway not having a synapse either in the thalamus or an analagous body such as the lateral geniculate nucleus.

INDEX